Mastering Basic Skills in Science

Preparing for Your Intermediate-Level Science Exam

SECOND EDITION

Donald D. Abramson

Former Assistant Principal
Supervisor of Biological and Physical Sciences
Martin Van Buren High School, New York City
and
Lecturer, Biology Department
Queens College, New York City

AMSCO

AMSCO SCHOOL PUBLICATIONS, INC.
315 Hudson Street   New York, N.Y. 10013

MASTERING BASIC SKILLS IN SCIENCE is dedicated
to my three young grandchildren, Dillon Zee, Logan Tyler,
and Jenna Rose, who have so profoundly changed my life.

The author wishes to acknowledge the contributions of
Herbert H. Gottlieb, Educator, Queens, New York,
in the preparation of this book.

Cover design/Text design: Merrill Haber
Cover photo: G. Biss/Masterfile
Illustrations: Hadel Studio
Composition: Publishing Synthesis, Ltd.

When ordering this book please specify: *either* **R 722 W** *or* MASTERING BASIC SKILLS IN SCIENCE
SECOND EDITION

Please visit our Website at: www.amscopub.com

ISBN 978-0-87720-047-5

To The Teacher:

One day a science teacher asked for a volunteer to come to the front of the classroom and read the temperature on a thermometer. To everyone's surprise, not a single hand was raised to show willingness to perform the task. So the teacher decided to appoint a "volunteer." This brave student struggled to read the thermometer while the rest of the students nervously looked on, each hoping that he or she would not be called on next to come forward. Finally, to the relief of everyone in the classroom, the teacher said, "Well, I'll give the reading for all of you this time," and signaled the student to sit down again.

Does this story sound familiar? Do your students know how to use measuring instruments, such as a metric ruler, a graduated cylinder, a triple beam balance, voltmeter, stopwatch, compass, or a thermometer? Have they ever watched and memorized laboratory procedures and techniques without fully understanding them or being able to take part in an experiment?

The "hands-on" activities in *Mastering Basic Skills in Science, Second Edition* are designed to (1) acquaint students with scientific procedures and (2) train them to use the measuring tools of science. This edition contains two new units (*Using the Microscope* and *Using Other Measuring Instruments and Techniques*) comprising 12 new chapters and skills. Chapters 7 and 8 are also new. Chapter 7 details the search for the cause of AIDS and the war against this disease. Chapter 8 presents the events leading to the identification of the West Nile virus in New York. Dr. Deborah Asnis, one of the scientists who worked on this puzzle, helped us with this reading passage.

Each worksheet is divided into sections. The *Title, Word Check*, and *Problem* introduce students to the skill that you want them to accomplish or learn to do well. The *Equipment* list and *Procedures* provide them with the necessary materials, instruments, and instructions needed during an experiment or activity. The *Test Your Understanding* and *Going Further* questions ask students to state their observations and conclusions during an experiment and encourage them to apply their newly gained knowledge, skills, and insights in solving other scientific problems.

The worksheets encourage students to acquire the following skills:

- Use the methods of science to answer questions and to solve problems.

- Identify, select, and safely use laboratory chemicals and equipment.

- Use measuring instruments to make metric measurements of length, mass, volume, density, and temperature

- New skills: using a microscope and using a spring scale to measure mass, measuring voltage and angular elevation

- Collect, record, organize, graph, and interpret data.

- Separate the parts of a mixture.

When they have mastered these skills, their science studies will become more meaningful and rewarding. The worksheets in this laboratory skills book will help them to change from being students who are simply *observing* science to those who are actually *doing* science.

The author hopes your students will enjoy learning science from this book.

Donald D. Abramson

Contents

Safety Rules for the Laboratory

When carrying out laboratory procedures, you must follow instructions and work safely. Safety rules to guide you when performing a laboratory lesson are listed below and are reviewed in the worksheets you will be using.

Before Starting a Laboratory Lesson

1. Know the locations of exits, fire extinguishers, fire blankets, and other safety equipment.

2. Know when and how to use safety equipment properly.

3. Read and understand all laboratory procedures in a lesson **before** you begin. Listen carefully to your teacher for additional instructions. Ask for help if you are not sure how to proceed.

4. Make sure loose, flowing clothing or long hair is securely fastened and out of the way.

During a Laboratory Lesson

1. Follow all directions exactly.

2. Work only when the teacher is present.

3. Concentrate on what you are doing. Do not move away from your work station unnecessarily. Be serious about your work.

4. Avoid eating or drinking in the laboratory. Do not use laboratory equipment as containers for food or drinks.

5. All accidents are to be reported immediately to the teacher.

Using Heat in a Laboratory Lesson

1. Wear safety goggles when heating materials and when working with chemicals.

2. When heating a material in a test tube, the mouth of the test tube should always be pointed away from you and from any other person.

3. Use heat-resistant glassware that is identified with a label such as Pyrex or Kimax.

4. Assume that glassware that has been heated recently is still hot to the touch. Handle it with a clamp or other holder.

5. If you burn yourself, run cold water over the burned area, and report the burn to your teacher immediately.

Working With Chemicals in a Laboratory Lesson

1. Use only the chemicals you have been instructed to use, and only in the way you are directed.

2. Chemicals are never to be tasted.

3. Chemicals are not to be touched. Dry chemicals are transferred with specific equipment such as a splint, spatula, or scoop.

4. Smelling a chemical is done by fanning the odor from a container to your nose.

5. If a chemical spills on your skin, wash it off with lots of cold water. Report all accidents to your teacher immediately.

6. Instructions on how to clean up chemical spills in your work area should be obtained from your teacher.

7. Broken glassware should be reported to the teacher. Then, the broken glassware should be cleaned up with a dustpan and brush and disposed of in a special container. Broken glassware is not to be picked up with your fingers.

8. Chemical containers that are missing labels should be reported to the teacher. Do not use chemicals that are in unlabeled containers.

Cleaning Up After a Laboratory Lesson

1. Leave enough time at the end of a laboratory lesson for cleaning up.

2. All students should be involved in the cleanup process.

3. Equipment should be taken apart and cleaned according to instructions.

4. Every item should be stored or replaced in its proper area.

5. Close all open drawers and cabinets to prevent someone from walking into them.

UNIT 1
USING
LABORATORY EQUIPMENT

C H A P T E R

1

Preventing and Controlling Fires in the Classroom

✓ WORD CHECK

fire prevention	to take measures to keep harmful fires from starting
substance	a material that makes up an object or thing
extinguish	to put out a fire (A fire *extinguisher* is a piece of equipment that puts out a small fire by releasing fire *extinguishing* substances.)
carbon dioxide	a type of gas found in fire extinguishers

PROBLEM: What should you know about fire prevention and control?

Fire in a science classroom can be either good or bad. When used carefully, fire is important for heating materials. When begun accidentally, fire can be very dangerous. Today, you will learn how your classroom is equipped to handle and control emergencies

1

involving fires. You will also learn about **fire prevention** and safe procedures to follow when using a flame.

PROCEDURES

Use complete sentences to answer the following questions. Base your answers on your observations, on information provided by your teacher, and on the *Safety Rules* at the front of this book.

CAUTION: Fire extinguisher demonstrations should be conducted only in open and safe areas and only by qualified or trained individuals in cooperation with personnel who are responsible for recharging the equipment.

A. Fire Extinguishers

1. Is there a fire **extinguisher** in the classroom?

    yes

    Where is it located?

    on the pillar by the board and next to the power shower

    Why is it important to know the location of the fire extinguisher *before* an accidental fire occurs?

    to save time during an emergency and get the fire extingusher quickly

2. Some fire extinguishers produce the gas **carbon dioxide**. Other types put out fire with a stream of water. Which type is in the classroom?

    carbon dioxide gas

3. The carbon dioxide fire extinguisher has several key parts: the pin, the horn, and the handle. Why is the pin important?

    to help avoid accedentally spraying the fire extinguisher

    Why should a person not pull the pin until reaching the fire?

    so they don't spray it and use up all the gas/water

4. The carbon dioxide comes out of the horn under a blast of pressure. Why would you have to be careful when aiming the horn at burning material in the classroom?

    so you dont accentally spray yourself or another person

5. Carbon dioxide gas is heavier than air. It helps to put out a fire by cutting off the supply of oxygen to the flame. Which part of the flame should the fire extinguisher be aimed at? Why?

base where the fire starts

Why might it be dangerous to use this type of extinguisher in a very small room containing a number of people?

your body absorbs CO_2 faster than air so you could suffocate

B. Fire Blankets

1. Is there a fire blanket in the classroom?

yes

Where is it located?

by the power shower

How do you get it out of its container?

open the velcro

2. How is a fire blanket used to put out a fire?

wrap it tightly around the fire

3. In what way do both the carbon dioxide fire extinguisher and the fire blanket put out a fire?

cut off the fire's source of oxygen

4. If a student's clothes caught fire, why would it be safer to use the fire blanket instead of the carbon dioxide fire extinguisher?

to avoid the person suffocate

C. Fire Exits

1. How many exits are there from the classroom?

_4_

Why is it important to have more than one exit?

_____

_____

_____

2. When using the exits in an emergency, why should you remain alert and listen for any additional teacher's instructions?

_____

_____

D. Fire Alarms

1. Is there a fire alarm signal box in the classroom?

_____

If not, where is the signal box that is closest to the classroom?

2. How is a fire signaled in the school?

_____

_____

3. If you have been directed to leave the building, why should you *not* stop to go to your locker?

_____

_____

4. After you have left the building, how do you know when it is safe to return?

_____

_____

5. Why are fire drills so important?

_____

_____

6. Why are false alarms so dangerous?

_____

_____

E. Fire Prevention

1. How are the matches and the striking surface of a book of matches positioned in relation to each other?

   _____

   _____

   Why do you think a book of matches is designed to have this feature?

   _____

   _____

2. If you were a fire inspector, what fire prevention procedures and measures would you look for when you inspect your building or classroom?

   _____

   _____

   _____

   _____

   _____

   _____

3. A student who has long, flowing hair has to use a flame to heat material. What fire prevention rule should this student follow?

   _____

   _____

4. Why must an accidental burn be reported to the teacher as quickly as possible?

   _____

   _____

   _____

TEST YOUR UNDERSTANDING

1. Garages are often equipped with a hanging pail of sand. If there were a small fire in a garage, how could a pail of sand be used to put the fire out?

   _____

   _____

How do sand and a fire blanket similarly put out a fire?

stop the fire from receiving oxygen

2. What do you think this statement means: The best time to fight a fire is before it starts.

know how to avoid fires so they don't happen in the first place.

3. If the fat in a pan of food you were cooking burst into flame, why would putting a cover over the pan put the fire out?

the fire can't receive oxygen

4. Conditions needed to start a fire include a fuel to burn, a temperature high enough to start the fuel burning, and a supply of oxygen. When a house is burning, fire fighters turn their water hoses onto the roofs of nearby buildings even though the buildings are not burning. How does this protect those buildings?

cools the building and the fire can't spread w/ fuel (wet roof)

GOING FURTHER

5. What emergency phone number should you dial if you want to report a fire?

911

6. Where is the fire alarm signal box nearest your home located?

I don't know

7. Why should you remain near the fire alarm box after you have sent in the signal?

I don't have one

8. Why is it important to have smoke detectors in your home?

to wake you up if there is a fire at night

9. Why should you check smoke detectors regularly?

   to make sure they work)

10. What does it mean when a smoke detector alarm begins to go on and off every few seconds?

    it's low on battery

11. Why should you record the date you install a new battery in a smoke detector?

    to know when to replace the battery

12. Some people disconnect the battery in their smoke detector because it goes off when they cook. What advice would you give to such individuals? Why?

    don't cook flamable food

13. Many homeowners have installed carbon monoxide detectors in their homes. The Consumer Product Safety Commission has recommended that these poison-gas-detecting devices be installed in two places in the home. One recommended place is next to the furnace. Where should the other carbon monoxide detector be placed? Why?

    in an attic/bedroom b/c people might go there alot

C H A P T E R

2

Acquiring Safe Working Habits

73

WORD CHECK

safe	to be free from the threat of danger, harm, or loss. (To prevent harm or loss to yourself, follow *safety* procedures in the classroom.)
goggles	protective glasses worn when you are experimenting.
chemicals	substances used in experiments

PROBLEM: What safety practices should be followed in the laboratory?

To make sure that you work **safely** in the laboratory, ask yourself before you begin an investigation, "What do I do if . . .?" This type of planning helps you in two ways. First, it makes you aware of problems that could occur. You can then take steps to prevent them from developing. Second, it prepares you to take the necessary, proper actions to handle a problem if one arises.

PROCEDURES

Think about the situations described in the questions below. Tell what you should do in each situation; base your answers on common sense and on what you have learned in earlier science studies. Also explain why you should take this action. Write the answers in complete sentences.

1. What do I do if my partner wants to begin an experiment before reading through all the procedures?

    read for yourself and make sure they do too

2. What do I do if another student tries to talk to my partner and me while the teacher is giving instructions?

   nicely ask them to stop

3. What do I do if my partner was absent when we learned where the fire extinguisher and fire blanket are and how to use them?

   let them know where everything is + how to use it

4. What do I do if I cut myself during an experiment and my partner asks me not to tell the teacher?

   tell the teacher

5. What do I do if I am not sure whether or not a piece of glassware that has been heated is still hot when I have to touch it?

   wait a few minutes and ask a friend or teacher

6. What do I do if I break some glassware?

   inform a teacher

7. What do I do if I can't find safety **goggles** to wear when I have to heat a **chemical** over a flame?

   don't heat the chemicals until you have goggles

8. What do I do if both my partner and I have to leave our workstation for a short time while we are heating something over a flame?

   turn off the burner

9. What do I do if my partner wants to mix a group of chemicals together "just to see what will happen"?

   nicely tell them not to

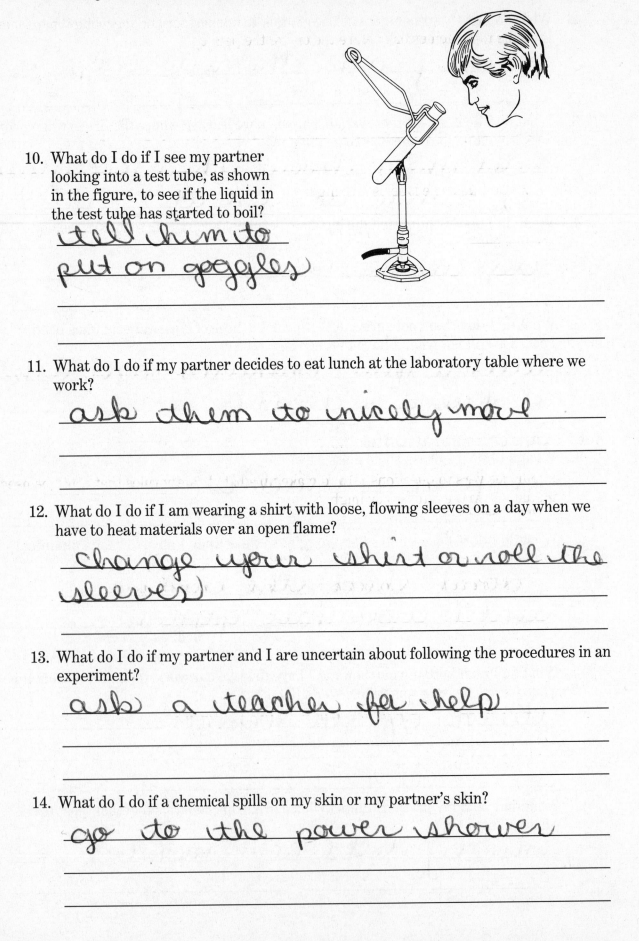

10. What do I do if I see my partner
    looking into a test tube, as shown
    in the figure, to see if the liquid in
    the test tube has started to boil?

    tell him to put on goggles

11. What do I do if my partner decides to eat lunch at the laboratory table where we
    work?

    ask them to nicely move

12. What do I do if I am wearing a shirt with loose, flowing sleeves on a day when we
    have to heat materials over an open flame?

    change your shirt or roll the sleeves)

13. What do I do if my partner and I are uncertain about following the procedures in an
    experiment?

    ask a teacher for help

14. What do I do if a chemical spills on my skin or my partner's skin?

    go to the power shower

15. What do I do if my partner wants to rush through an experiment so that there will be time to do homework before the end of the period?

tell them that homework is for home and that we need to make sure we do our experiment properly

16. What do I do if the label is missing from a bottle containing a chemical that is going to be used in an experiment?

don't use it

TEST YOUR UNDERSTANDING

Use your answers to questions 1-16 to make up a list of safety rules that could be used by other students in the future. You might want to start each rule with terms such as *Always* or *Never*. The first two rules have been done for you as examples.

1. *Always read all procedures through before starting an experiment.*

2. *Always listen carefully when the teacher is giving instructions.*

3. _____

4. _____

5. _____

6. _____

7. _____

8. _____

_____

9. _____

_____

10. _____

_____

11. _____

_____

12. _____

_____

13. _____

_____

14. _____

_____

15. _____

_____

16. _____

_____

GOING FURTHER

Many of the safety rules followed in the laboratory should also be followed at home. How should you safely act in each of these situations?

17. A container of pills in the medicine cabinet has no label.

_____

_____

18. You have just boiled an egg in a pot on the stove. How can you safely pick up the pot to pour off the water into the sink?

_____

_____

_____

3

Selecting Suitable Equipment

☑ WORD CHECK

flask	a container with a long, narrow neck used to hold liquids
stopwatch	a device used to measure time
thistle tube	a funnel with a bulging top

PROBLEM: What equipment is used when working with liquids?

As you study science, you will often observe, describe, and interpret investigations demonstrated by your teacher. At other times, you will carry out investigations yourself. In either situation, special types of laboratory equipment will be needed. Your work today will acquaint you with some of the equipment used when working with liquids. You will also learn how to choose the appropriate equipment for the procedure you wish to carry out.

PROCEDURES

Your teacher has arranged three groups of equipment for you to inspect. The equipment is also shown in illustrations that are grouped as either A, B, or C. Examine each piece of equipment in the groups on display. As you do so, find the matching object in the illustration and place a check in the blank next to its name to indicate that you have studied the particular piece of equipment. Then answer the questions and carry out the instructions given.

Group A   Containers for Liquids

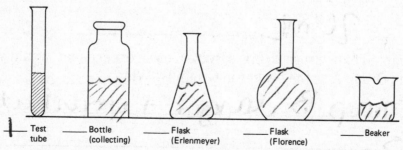

___ Test tube   ___ Bottle (collecting)   ___ Flask (Erlenmeyer)   ___ Flask (Florence)   ___ Beaker

1. In the drawing of the test tube, the slanted lines (/ / / / /) show that liquid is present in the test tube. The surface of the liquid is indicated by the horizontal, wavy line (～～～). Add marks to the other drawings in Group A to show that each piece of equipment holds a liquid.

2. Notice that the edge of the beaker has a *lip*, a small spout such as those of some pitchers. What does the lip tell you about one of the purposes of a beaker?

   it is used for liquids + pouring

Group B   Equipment for Measuring Liquids

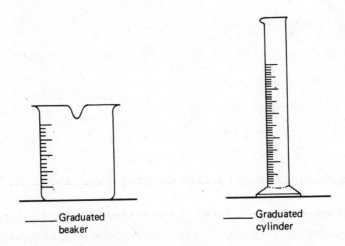

_____ Graduated
beaker

_____ Graduated
cylinder

1. How does the beaker in Group B differ from the beaker in Group A?

   Group B is larger + has marked measurements

   Each of the lines on the beaker and the cylinder is known as a *graduation*. Notice that the graduations are evenly spaced and some are numbered. The graduations enable you to measure the amount of liquid you put into or pour out of these pieces of equipment.

2. Look at the graduated beaker. Near the top of the beaker is a number followed by **mL**. The **mL** stands for a unit by which liquids are measured in science. It is called the milliliter. What is the largest amount of liquid you can measure with the beaker on display in Group B?

   70mL

3. If you wanted to measure 25 mL of water for an experiment, would you use the beaker in Group A or the beaker in Group B? Why?

   Group B-larger + marked by 5's

4. Inspect the graduations on the cylinder. What is the largest amount of liquid you can measure with this cylinder?

    20 mL   25mL

Group C   Equipment for Transferring Liquids

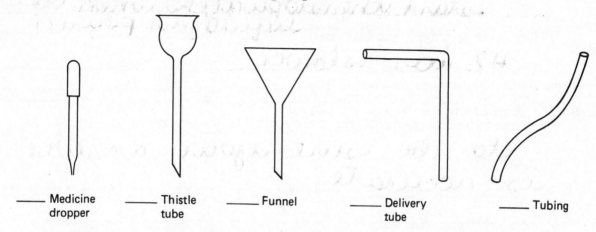

___ Medicine   ___ Thistle   ___ Funnel   ___ Delivery   ___ Tubing
     dropper   tube   tube

1. Why would you sometimes use a medicine dropper instead of a beaker or a graduated cylinder to transfer a liquid from one container to another?

    to get more precise amounts

2. Why can't you accurately tell the amount of liquid you transfer with a dropper?

    it varies everytime and there aren't measurment markings

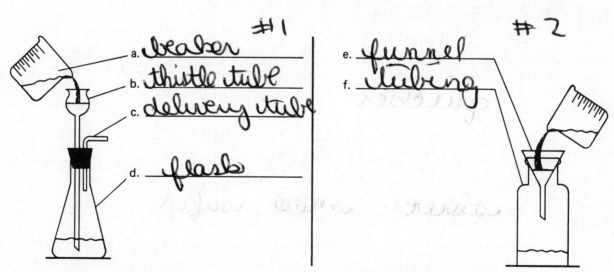

#1
a. beaker
b. thistle tube
c. delivery tube
d. flask

#2
e. funnel
f. tubing

3. Your teacher will demonstrate two methods of transferring liquids, using the equipment shown in the figures above. Before the demonstration begins, write the name of the lettered pieces of equipment in the spaces provided.

4. If equal amounts of water are added at the same time to the containers in both setups, which container do you predict will fill more quickly?

   II/

   How would you use a stopwatch to check your prediction?

   start the stopwatch when the

   Perform the activity. What happened? *liquid is poured*

   #2 was slower

5. Why must you use equal amounts of water in this demonstration?

   to be sure your answer is accurate

6. Compare how far the **thistle tube** and the funnel reach into the containers. What advantages does the longer stem of the thistle tube have over the shorter stem of the funnel?

   the thistle tube is slower and therefore usually safer

TEST YOUR UNDERSTANDING

1. Why is it sometimes an advantage to use a funnel when transferring a liquid into a container?

   quicker

2. Why is it sometimes an advantage to use a thistle tube when transferring a liquid into a container?

   slower - more safer

3. Why is it sometimes an advantage to use a medicine dropper when transferring a liquid into a container?

   slower and you have more controll over the ammount you are transferring

The following figure shows equipment that has been set up to produce a gas.

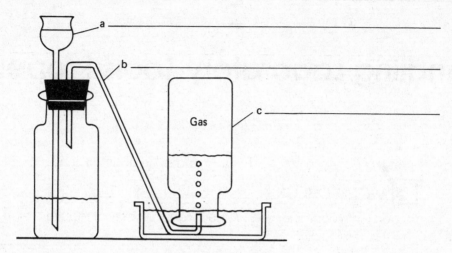

4. Name each of the lettered pieces of equipment.

    a _____

    b _____

    c _____

5. Which of the lettered pieces of equipment is being used to transfer a liquid?

    _____

6. Which of the lettered pieces of equipment is being used as a container?

    _____

7. Which of the lettered pieces of equipment is being used to make measurements?

    _____

4

Handling Laboratory Tools Properly

☑ WORD CHECK

crucible	a porcelain container used to heat substances that require a high degree of heat to melt
hot plate	a piece of equipment used to heat substances without using an open flame

PROBLEM: How is equipment used to hold objects and heat materials?

You have learned about equipment commonly used when working with liquids. Today, you will observe and use other types of equipment with different functions.

PROCEDURES

After you have handled and inspected the equipment your teacher has set up, find the drawing of each piece of equipment in Groups D, E, and F below. Place a check in the blank next to the name of the object to show that you have studied it.

Group D   Equipment for Picking Up or Holding Other Equipment

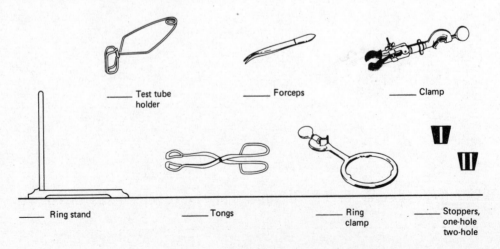

_____ Test tube holder   _____ Forceps   _____ Clamp

_____ Ring stand   _____ Tongs   _____ Ring clamp   _____ Stoppers, one-hole two-hole

1. Why is it better to attach equipment to a ring stand than to let the equipment stand on the tabletop by itself?

    _____

    _____

2. Which pieces of equipment in Group D are used to attach other pieces of equipment to a stand?

    _____

    _____

3. Handle a test tube holder to see how it works. Then, pick up a test tube with the holder. When holding a test tube, why should you keep your fingers near the base of the holder and *not* on the pressure points at the widest part of the holder?

    _____

    _____

    _____

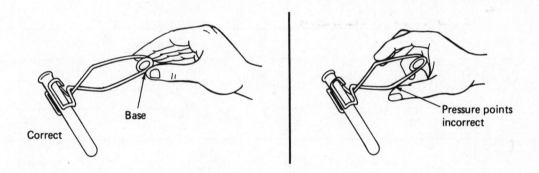

Group E   Equipment for Heating Substances

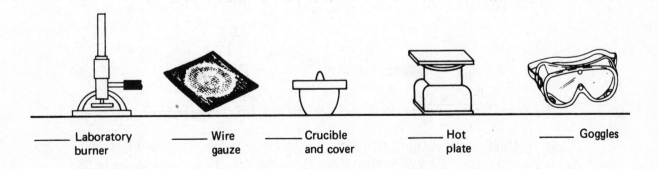

_____ Laboratory   _____ Wire   _____ Crucible   _____ Hot   _____ Goggles
burner   gauze   and cover   plate

1. Which piece of equipment in Group E is most directly related to your safety?

    _____

2. What rule on the list of safety rules in the front of the book should you follow when heating a material in either a test tube or a beaker?

_____

_____

3. When you heat a liquid in a flask or beaker over a flame, why is it an advantage to place a wire gauze between the bottom of the glassware and the flame?

_____

_____

In science laboratories, it is often necessary to cut glass tubing into shorter lengths or to bend glass tubing into different shapes. Your teacher will demonstrate these techniques.

4. How is a file used to aid in cutting glass tubing?

_____

_____

5. How does the end of a piece of newly cut glass tubing change when it is rotated in the burner flame?

_____

_____

Why is this procedure called *fire-polishing*?

_____

_____

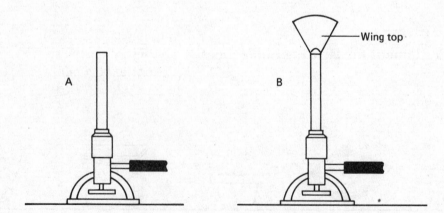

6. Examine the figures of the above two laboratory burners. The one on the right, B, has had a wing top fitted to it. How does the wing top change the shape of the flame?

_____

Draw the flames above each of the burners to show their differences.

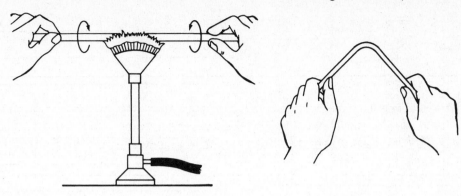

7. Your teacher will use a burner flame with and without a wing top to bend glass tubing. How does the bend differ when a wing top is used?

_____

_____

8. When you are bending glass to make a delivery tube, why is it an advantage to use the wing top on the burner?

_____

_____

_____

TEST YOUR UNDERSTANDING

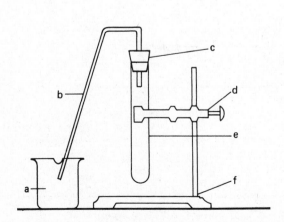

The following figures show various pieces of laboratory equipment set up for carrying out different experiments. For each figure, do the following three things:

1. Name each piece of equipment.

	Name	Use
a.	_____	___
b.	_____	___
c.	_____	___
d.	_____	___
e.	_____	___
f.	_____	___

a. _____ _____

b. _____ _____

c. _____ _____

d. _____ _____

e. _____ _____

f. _____ _____

g. _____ _____

h. _____ _____

2. After the name of the item, tell its use by writing the letter of the group to which it belongs:

   Group A   Equipment used as containers for liquids

   Group B   Equipment used for measuring liquids

   Group C   Equipment used for transferring liquids

   Group D   Equipment used for picking up and holding

   Group E   Equipment used for heating

3. Show in the figures that each container is half-filled with a liquid.

GOING FURTHER

4. Choose one of the setups in the figure shown in "Test Your Understanding." Obtain the necessary pieces of equipment and set them up as indicated in the figure. **CAUTION:** Do not insert delivery tubes or thistle tubes into a stopper. A special technique must be used for these procedures to avoid the possibility of a serious cut. Your teacher will provide the stoppers with the necessary equipment inserted. If they are not available, ask your teacher for help. Do not attempt to remove glass tubing from a stopper. This also requires a special technique.

C H A P T E R

5

Using a Laboratory Burner

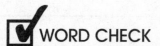 WORD CHECK

| **Bunsen burner** | a heating device that mixes gas and air to provide a very hot flame for laboratory use |

PROBLEM: How is a laboratory burner used?

A burner is a standard piece of equipment for heating materials in the laboratory. Today, you will learn the following: (A) the parts of a burner, (B) how to use a burner safely, and (C) how it works.

EQUIPMENT

Check off the items to make sure you have the following equipment. If any item is missing, obtain it from your teacher. Depending upon the availability of equipment and facilities, this activity may be performed as a demonstration rather than as a laboratory lesson.

___ laboratory burner   ___ test-tube holder   ___two test tubes

PROCEDURES

Your laboratory burner is most likely a **Bunsen burner**, as shown in the figure below. Other types of burners are used in some classrooms. If the one you are observing is different, draw your burner in the space next to the figure of the Bunsen burner.

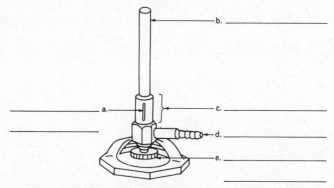

Part A

1. Label the parts of the figure indicated by the lettered lines and according to information provided by your teacher.

2. Where does the gas enter the burner?

   _____

   _____

3. How is the amount of gas that enters the burner controlled?

   _____

   _____

4. Where does air enter the burner?

   _____

   _____

5. How can you control the amount of air that mixes with the gas?

   _____

   _____

Part B

Recall, and also describe, the safety rules to be followed in each of the following situations:

1. To use the burner, you are about to turn on the gas and light the gas with a match. Which must be done first?

   _____

2. How would you use a test-tube holder when holding a test tube in a flame?

   _____

   _____

3. What *two* points concerning eye safety must you remember when heating a test tube?

   _____

   _____

Part C

1. Turn the burner's collar to close the air intake openings, and then light the burner. What color is the flame?

   _____

2. Hold the bottom of a test tube in this flame for a brief time. How does the bottom of the test tube look when it is removed from the flame?

_____

_____

3. Open the air intake collar while the burner is still on. What color is the flame now?

_____

_____

4. Hold a second test tube in this flame for a brief time, and then examine its bottom. How does the bottom of the test tube compare in appearance with the bottom of the first test tube?

_____

_____

5. When a burner is working most efficiently, the gas burns completely in the hottest possible flame, and no remnants (ashes or soot) are produced. What color was the hottest flame?

_____

What evidence was there that this color flame was the hottest?

_____

_____

_____

TEST YOUR UNDERSTANDING

Look closely at the blue flame. You should observe an inner and outer part to the flame. These different parts are called *cones*.

1. Draw the cones on the figure of the Bunsen burner Label the inner and outer cones. How do these different parts of a flame compare? Your teacher will help you arrive at this answer by using the setup shown in the figure on the next page. Note how the match is positioned with a pin through the wood so that the head of the match will be in the center of the flame when the burner is lit.

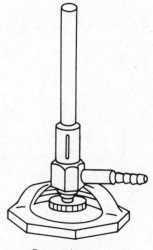

Bunsen burner

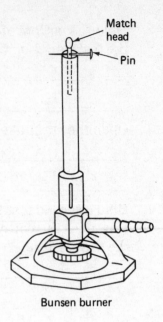

Bunsen burner

2. What do you predict will happen to the match when the burner is lit?

_____

_____

3. What happens?

_____

_____

_____

4. Why did this happen?

_____

_____

_____

GOING FURTHER

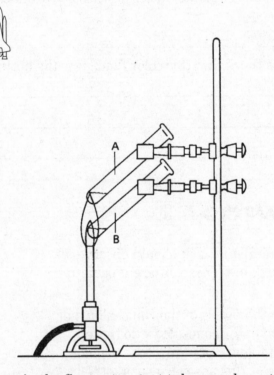

5. In the setup shown in the figure, two test tubes, each containing the same amount of water, are heated at the same time but in different parts of the flame. In which test tube, A or B, would you expect the water to begin to boil first? Why?

_____

_____

_____

_____

6. Why is it important that the test tubes in this demonstration contain equal amounts of water?

_____

_____

_____

7. Predict what would happen to the different lettered parts of the wooden splint shown in the figure if it were placed briefly in the flame of a Bunsen burner. Why?

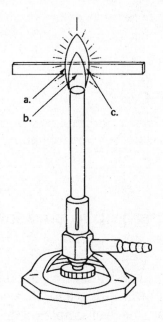

_____

_____

_____

CHAPTER

6

Practicing Laboratory Techniques

 WORD CHECK

| AIDS virus | a virus (acquired immunodeficiency syndrome) that destroys one of the body's important defenses against diseases |

PROBLEM: What precautions help to prevent contamination of chemicals?

Scientists often use chemicals called **indicators** in their laboratory investigations. Indicators, as their name suggests, indicate, or show, that something is present by changing color. Some indicators detect the presence of simple sugars, starch, carbon dioxide, proteins, and vitamin C. Today, you will use an indicator that detects the presence of substances known as **bases**. Examples of bases include soap, lye, ammonia water, and milk. The base indicator that you will use is a liquid called phenolphthalein (fee-nul-*thal*-ee-un).

EQUIPMENT

Make sure you have at your work station all the equipment and materials that are listed below. Place a mark next to each item as you check that it is present. If an item is missing, obtain it from your teacher.

_____ 2 test tubes   _____ dropper bottle containing a base

_____ test-tube rack   _____ dropper bottle with phenolphthalein

_____ small funnel

PROCEDURES

1. What happens when your teacher places drops of phenolphthalein into test tubes that contain different types of bases?

_____

Why should this also be done with a test tube of water?

_____

_____

2. Place two test tubes on a rack. Place a small funnel in one of the test tubes on the rack. Fill this test tube about halfway with the base from the dropper bottle. Do this by pouring the base from the bottle rather than by using the bottle's dropper. Why do you use a funnel when pouring a liquid directly into a test tube?

_____

_____

3. How do the base and the phenolphthalein indicator compare in appearance?

_____

_____

4. Use the medicine dropper to add just one drop of the phenolphthalein indicator to the base in the test tube. Describe what happens.

_____

_____

5. Now use the same medicine dropper to add exactly 5 drops of the base to the second test tube. NOTE: Before you do this, squeeze the rubber bulb of the dropper a number of times to get rid of any of the phenolphthalein indicator that may be left in it. When you are sure that all the indicator has been removed from the medicine dropper, fill the dropper from the bottle of base. Describe what happens.

_____

_____

6. What explanation can you offer for what happened when you began to fill the dropper with the base?

_____

_____

7. What procedural changes should you make in step 5 to avoid the same thing happening in the future?

_____

_____

8. The liquid indicator that was still in the dropper **contaminated** the base in the bottle. What do you think the term *contaminated* means?

_____

_____

_____

9. Now that the base in the bottle has been contaminated, should the bottle be returned to the supply shelf for use in future work? Why?

_____

_____

_____

10. Why should you always report any possible contamination of a chemical to your teacher?

_____

_____

_____

_____

TEST YOUR UNDERSTANDING

1. When a large jet airplane is repaired, the mechanics remove all the plane's fuel from its tanks. This safety procedure reduces the chances of an accidental fire. When the repairs are completed, the plane is refueled, but the original fuel is not returned to the tanks. Why is new fuel used instead of the original fuel?

_____

_____

_____

2. What should you do if your partner pours out more of a liquid than is needed for an experiment and is about to pour the extra liquid back into the original supply bottle? Why?

_____

_____

_____

3. What should you do if your partner uses a scoop to remove starch from a bottle and then is about to use the same scoop to take sugar from another bottle? Why?

_____

_____

_____

_____

GOING FURTHER

4. If the stopper of a bottle containing a chemical is placed on the laboratory tabletop, it may become contaminated by materials on the tabletop. When replaced in the bottle, chemicals on the stopper may mix with the contents of the bottle. Describe the technique (shown in the figure) your teacher uses to prevent contamination of this kind.

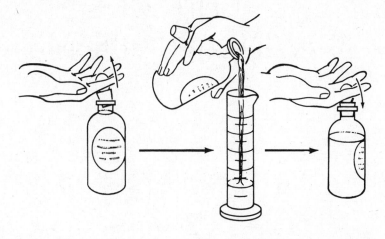

_____

_____

_____

_____

5. A group of individuals who are in great risk of developing the disease **AIDS** are drug abusers who inject themselves with needles shared by other drug abusers. Why is the virus that causes this disease so much more likely to be spread among drug abusers?

_____

_____

_____

UNIT 2
SOLVING PROBLEMS

CHAPTER

7

Reading For Facts:
Applying Facts to Seek Solutions

 WORD CHECK

viruses	extremely small particles that can cause diseases by entering cells, multiplying, and destroying the cells
host cells	cells that, when invaded by viruses, are forced to produce new viruses
helper T cells	a specialized type of white blood cell that helps to destroy viruses
antibodies	chemicals in the blood that work with T cells to destroy viruses
immunity	a condition of the body that occurs when it has sufficiently high levels of antibodies to prevent a disease from developing
HIV	(human immunodeficiency virus) the virus that cripples the body's immune system by destroying helper T cells
AIDS	(acquired immunodeficiency syndrome) the group of diseases or conditions resulting from an invasion of the body by the HIV virus

PROBLEM: What can you learn from reading science stories?

Almost daily, newspapers and magazines give us a great deal of factual, science information. It is not enough however, for you simply to be told the facts. You must also understand the material for it to be useful. In the story you will read today, you will learn about a terrible disease. Then you will be asked to use the information presented to identify possible solutions that could slow or stop the spread of this disease.

PROCEDURES

Read the following story. Then answer the questions at the end.

A Rare Puzzle

The Puzzle

In the early 1980s, public health doctors in San Francisco, California, began to notice an unusual increase in the number of patients who had developed a deadly form of cancer, Karposi's sarcoma. What puzzled the doctors about this disease is that it occurs only rarely. So rarely, in fact, that most physicians never see even one case in their entire career.

Soon, the doctors collected more facts about this disease. Almost all the victims were young males in their late teens or early twenties, and they were all members of San Francisco's gay community. In addition to this type of cancer, there were reports that an equally deadly form of pneumonia was spreading among intravenous (IV) drug abusers who shared blood-contaminated syringes and needles with other drug abusers. More new cases of Karposi's sarcoma and pneumonia were reported among hemophiliacs who regularly received blood transfusions. Cases of these diseases also occurred in patients who had received organ transplants and were given medicines to suppress their immune systems.

What was going on? What was causing these illnesses? What did all these patients have in common? The doctors gave their patients blood tests to try to find answers to some of these questions.

The Answers Begin to Emerge

Karposi's sarcoma is an "opportunistic" disease. Opportunistic diseases develop in people whose immune systems are damaged, and whose bodies are unable to resist invasion by diseases. To understand more about opportunistic diseases, you need to know about the body's immune system.

In your bloodstream, there are two main fighters against disease. First, there are the chemicals called **antibodies**. Antibodies target and destroy disease-causing organisms or the toxins (poisons) produced by these organisms. Second, there are the white blood cells known as **helper T Cells**. These cells destroy germs and assist in forming additional antibodies when infectious agents, such as bacteria or viruses, enter the body.

Now, you know something about the people who developed Karposi's sarcoma. In addition, you know a bit about the workings of the immune system. So, you can understand why the doctors were anxious to see the results of the blood tests they gave to their patients, in search of a common cause for these rare diseases. And there it was! Every patient had strikingly low levels of helper T cells. Something was destroying these cells faster than they could be produced, thereby leaving the body open to infection. But what was doing this?

The Culprit

In 1984, groundbreaking news came from a research laboratory in France. It's a **virus**! Helper T cells were being invaded and destroyed by **HIV** (human immunodeficiency virus). Once HIV is in the helper T cell, it begins to multiply and eventually kills the cell. Illnesses such as Karposi's sarcoma were successful in attacking the body after a person's immune system had been weakened or overcome by HIV. The progression of the disease can be described as a race, or competition, between HIV and helper T cells. If HIV multiplies faster than helper T cells, an infected individual would become sicker and sicker and perhaps die. The different diseases resulting from HIV are known collectively as *HIV/AIDS* (human immunodeficiency virus/acquired immunodeficiency syndrome).

The War Against HIV/AIDS: The Understanding

Now that the cause of HIV/AIDS had been identified as a virus, two more questions arose. First, how is the virus transmitted from person to person? Second, what can be done to stop the spread of this disease? Interestingly, answering the first question led to the answer of the second question as well. While there are medicines to treat HIV/AIDS and help patients live longer, stopping the spread of this virus is particularly important because there is no cure for HIV/AIDS.

The War Against HIV/AIDS: The Attack

One by one, researchers eliminated the possible methods by which the virus could be transmitted. The virus does not spread in the air, in water, in food, or by casual contact. It now became obvious, and the data confirmed this, the virus travels from person to person in body fluids such as blood or during intimate contact.

To stop transmission of the disease, "at-risk" people had to be educated and programs established to increase awareness of the dangers. Young people in particular are being advised of the value of abstaining from intimate contacts or encouraged to take precautions.

The War Against HIV/AIDS: An Update

By the end of 2000, an estimated 36 million people worldwide were HIV/AIDS infected and 21 million people had died. In the United States, some 650,000 to 900,000 individuals were infected. Scientists reported that 430,441 people had died by the end of 1999. HIV/AIDS is now the fifth leading cause of death among young people in the United States.

In January 2001, a major news program painted a new and alarming picture of the occurrence of HIV/AIDS in the United States. The broadcast reported that, after several years of a decrease in the number of cases, doctors were seeing an increase in cases among members of the gay community. Public health officials theorize that this increase is a result of the false belief that, with the introduction of combinations of new medicines (cocktails), the use of protection during intimate relations was no longer necessary.

The War Against HIV/AIDS: The Future

Information from the Centers for Disease Control and Prevention (CDC) provides the following data about HIV/AIDS-related deaths in the United States.

YEAR	HIV/AIDS-Related Deaths
1996	37,739
1997	21,850 (down 42% compared with the previous year)
1998	17,840 (down 18% compared with the previous year)

These results were encouraging. Scientists continue the search for a vaccine or a cure for this disease.

TEST YOUR UNDERSTANDING

(Write the letter of the correct answer in the space.)

1. The letters HIV describe _____
    a. a disease   c. a type of virus
    b. a group of diseases   d. a type of white blood cell

2. When doctors in San Francisco first became aware of how widespread HIV/AIDS was becoming, they discovered that the infected males were primarily in what age range? _____
    a. 15-24   c. 35-44
    b. 25-34   d. 45-54

3. The virus that causes HIV/AIDS is transmitted by _____.
    a. food that has been improperly handled
    b. bottled water that has been contaminated
    c. polluted air from automobile exhausts
    d. an unprotected exchange of body fluids

4. HIV/AIDS cannot be cured with antibiotics such as penicillin because the virus that is responsible only reproduces _____.
    a. inside invaded cells   c. in warm areas
    b. on dry surfaces   d. where there is enough food

5. Drug abusers are more likely to be infected with the virus that causes HIV/AIDS when their lifestyles include _____.
    a. diets lacking in the proper vitamins and minerals
    b. sharing syringes used to inject drugs
    c. living in crowded conditions
    d. rare doctor visits

6. In this country, the number of reported cases of HIV/AIDS in 1997 was between _____.

   a. 100,000 and 300,000   c. 600,000 and 900,000
   b. 300,000 and 600,000   d. 900,000 and 1,200,000

7. The earliest reported cases of HIV/AIDS in this country was in _____.

   a. 1961   c. 1981
   b. 1971   d. 1991

8. What information from the table on page 35, which gives statistics for 1996 to 1998, indicates that in the United States the war against HIV/AIDS had been succeeding? _____

   a. The number of deaths each year has been increasing.
   b. The number of deaths each year has been decreasing.
   c. The number of deaths each year has remained about the same.
   d. The number of deaths each year has increased then decreased.

9. In section C, the term "culprit" is used. Which one of the following terms best fits that description? _____

   a. helper T cells   c. a virus
   b. addicting drugs   d. antibodies

10. According to the January 2001 report, why was there an increase in the number of cases of HIV/AIDS in the United States?

   a. The new drugs do not work as well as the older drugs.
   b. People believe that the combinations of new medicines work so well that they no longer have to use protection.
   c. There was no increase, the report was wrong.
   d. The HIV/AIDS virus has become resistant to drug therapy.

GOING FURTHER

11. Why must health-care professionals, such as doctors, dentists, nurses, and Emergency Medical Technicians, be particularly careful to wear gloves when tending patients who are bleeding?

    HIV/AIDS can be passed by blood

12. Why in the early 1980s did some doctors not immediately recognize the symptoms of Karposi's sarcoma in their patients?

    It was such a rare disease that may doctors never saw a case of it during their career

13. If a doctor suspects a patient has an infected appendix, why would a blood test be an appropriate procedure?

They can see the # of helper T cells + antibodies

14. Biology students once used their own blood to prepare slides for examination under their microscopes. Why are commercially prepared slides now being used instead?

to prevent accidentally spreading diseases + viruses through blood + to see different types of blood

15. What can young people do to protect themselves from infection with HIV?

always use protection during sex + don't take drugs

16. When donors contribute blood, it is now "HIV screened" before it is used in transfusion. What does this mean and why is it being done?

blood is checked before being used in transfusion so that diseases + viruses aren't spread

CHAPTER

8

Identifying Experimental Procedures

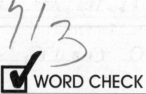

☑ WORD CHECK

symptoms	the characteristic signs of a disease
diagnosis	the identification of a disease based on the symptoms
epidemic	a sudden and rapid spread of a disease among many individuals at the same time

PROBLEM: How do scientists find answers to questions?

On September 11, 2000, the United States Government Accounting Office issued a report to a group of senators and congressmen. The report was in response to the legislators' concern over the sudden appearance in the New York City area of a disease never before seen in this part of the world. Seven people had died and dozens more had been made sick. All had the same **symptoms** and were **diagnosed** as having the same disease. Doctors were worried. Was this the beginning of an epidemic?

PROCEDURES

In your work today, you will read a scientific mystery based on the government's report. The story and the questions that follow will give you an example of how scientists work to solve problems and answer questions.

The Buzz Around New York

The First Puzzle

In the fall of 1999 at Flushing Hospital, Queens, New York, Dr. Deborah Asnis noticed something unusual about the seriously ill patients she was admitting. Most of her patients were elderly and had symptoms that included inflammation of the brain (encephalitis) and complications such as stomach complaints, muscle weakness, and breathing difficulties. Dr. Asnis had never had so many patients with encephalitis in just a two-week period. Also, it was unusual for encephalitis to strike elderly patients. In addition, encephalitis rarely causes muscle weakness. Because of this complaint, Dr. Asnis called the New York City Department of Health for help. The Health Department arranged for the patients' tissue samples to be sent to specialized state and federal laboratories.

The scientists in the laboratories diagnosed the disease as St. Louis encephalitis. This illness is transmitted by the bite of infected mosquitos. St. Louis encephalitis occurs in about 30 people annually throughout the United States, as well as in New York State, but has never been seen in New York City. The scientists continued to test specimens and realized that the virus was not St. Louis virus but *West Nile virus*. This virus was first isolated in Africa, and has been discovered in the Middle East, Asia, and parts of Europe. This was the first time that this virus had ever appeared in the Western Hemisphere. How did it get here? The answer may be in the second puzzle of that summer.

The Second Puzzle

In New York City, not only were patients becoming sick from this strange virus, but the birds were also dying! Large numbers of dead birds were being reported to biologists who specialize in the study of birds. Throughout the area, crows were found dead. Birds in zoos as well as in the wild were dying. What was causing this? Was this due to increased air pollution, or perhaps contamination of the city's water supply, or to some other as yet unsuspected cause?

At first no one connected the two puzzles. The human illness and the dying birds were viewed as separate events. Gradually, as an increasing number of laboratories became involved in testing human, bird, and mosquito tissue samples, the linkage became clear.

The birds were infected with West Nile virus. How did the birds become infected? Mosquitoes carrying the West Nile virus were biting birds and thus transmitting the virus to birds. These infected birds served as hosts for the virus. Then when the infected birds were bitten by uninfected mosquitoes, the virus spread. Eventually the virus killed the infected birds. The disease is transmitted to humans by the bite of an infected mosquito.

To prevent the spread of the West Nile virus, mosquito breeding areas were sprayed with insecticide. People were advised to use insect repellents. Insect repellent was distributed to the population at fire houses around the city. Citizens were told to look for and remove puddles of water (where mosquitoes lay their eggs). They were also told to wear protective clothing.

Conclusions

Our world is filled with questions to be answered and problems to be solved as in the story of HIV/AIDS (see chapter 7) and West Nile virus. In science, seeking solutions to problems is accomplished by using a series of steps known as the *scientific method*.

1. First, a specific problem is identified. (In chapter 7, for example, you read about HIV/AIDS. The question, "How is this disease transmitted?" was the starting point for the research.)
2. Based on the few facts available at first, an "educated" guess or *hypothesis* is formed. ("HIV/AIDS seems to be transmitted by body fluids such as blood.")
3. To determine if a hypothesis is correct, basic skills such as observing, measuring, and experimenting are used to *collect data*. ("Originally, HIV/AIDS was occurring mainly among homosexuals, IV drug abusers, hemophiliacs, and organ transplant recipients.")
4. Using the information collected, conclusions are arrived at that answer the original problem. ("HIV/AIDS is transmitted from one person to another in body fluids.")
5. Finally, the information gained is applied to raising new questions (What's in the blood that causes HIV/AIDS?") or using the information ("How can the knowledge be used to stop the spread of HIV/AIDS?")

TESTING YOUR UNDERSTANDING

For each of the following, match the letter of the statement to the step in the scientific method it represents.

1. The problem .... __c__   a. Prevention of West Nile virus depends on spraying chemicals to kill mosquitoes, but the question of how the virus got to this country remains unanswered.

2. The hypothesis .... __c__   b. Laboratory examination was done of virus-containing tissue from infected mosquitoes, birds, and humans and then compared with St. Louis virus samples.

3. The evidence .... __b d__   c. The basic question is what is causing this disease and does "it" also cause St. Louis encephalitis?

4. The conclusion ..... __e__   d. It's entirely possible that this disease and the one caused by the St. Louis virus are two different diseases.

5. The application .... __a__   e. The laboratory results indicated that the St. Louis virus and the West Nile virus are two different viruses producing two different diseases.

GOING FURTHER

6. Some of the different types of birds found in the New York City area spend the winter in warmer parts of the country. Why would this fact be a cause for concern for the public health officials in those parts of the nation?

   if the infected birds migrate to other parts of the nation + are bitten by mosquitos that then bite humans the disease might spread

7. When a mystery disease such as the West Nile virus disease suddenly appears in a community, what is the most appropriate method to arrive at a diagnosis? ____ (Write the letter of the correct choice in the answer space.)

   a. taking a vote among the doctors involved

   b. taking a vote among both the doctors and their patients

   c. taking a vote among doctors and public health officials

   d. collecting the evidence

8. During the most recent outbreak of the West Nile virus, public health officials had caged chickens placed in areas where there were no cases of the disease. Why did they do this?

    So that the chickens wouldn't become infected + produce eggs that would infect humans if consu consumed

9. There is no cure for the West Nile virus, although patients do receive treatment for the symptoms. If you were the public health official involved, how would you propose to fight the spread of this disease?

    I think that the steps the public health officials took to fight the spread of the West Nile disease was good.

C H A P T E R
9

Analyzing Experimental Procedures

✓ WORD CHECK

organism	a living thing
bacteria	some one-celled organisms that often cause disease
petri dish	a special flat container with a cover that is often used for growing bacteria in the laboratory
agar	the jellylike material that contains nutrients for the growth of bacteria studied in laboratories
spore	the reproductive cell of a mold (With proper light and food, a *spore* develops into a new mold.)
mold	a plantlike organism that cannot make its own food, it lives on other living things or on materials such as food and leather

PROBLEM: How is a controlled experiment planned and carried out?

In an earlier lesson, you analyzed a study carried out in the field. Today, you will read about a famous experiment that led to the discovery of a medicine that has saved countless lives. This experiment provides a good example of the way scientists carry out investigations in the laboratory.

PROCEDURES

Read the following description of the circumstances that led to the experiment and how the experiment was carried out. Then answer the questions under Analyzing the Experiment.

How the Experiment Came About

One day in 1928, Alexander Fleming, a British scientist, was working in his laboratory. He was

Petri dish with colonies of bacteria

42

studying harmful **bacteria**, which he grew on **agar** in **petri dishes**, as shown in the figure. Although Fleming did not know it, some **spores** of a green **mold** fell from the air onto the agar in one uncovered petri dish. There, the spores began to grow and reproduce.

When Fleming examined this petri dish some time later, he saw that mold was growing in the dish, along with the bacteria. But, to his surprise, he noticed clear spaces around the mold, where bacteria had once been growing. Checking further, Fleming found that all the bacteria had died in these clear areas around the mold. Instead of throwing out the ruined petri dish, Fleming decided to try to find out why the bacteria near the green mold had died.

Fleming's Experiment

A. Fleming began his work by asking a question: Is the green mold the reason the bacteria died?

B. Fleming knew that any number of things could have caused the bacteria to die, but he guessed that the green mold was responsible.

C. To see if his guess was correct, Fleming added green mold spores to other petri dishes in which the bacteria were growing.

D. He set aside an equal number of petri dishes in which bacteria were growing, but to these dishes he did *not* add mold spores.

E. After some time had passed, Fleming inspected both sets of petri dishes. This is what he saw: The petri dishes to which the mold spores had been added had clear spaces around the mold. All the bacteria in these clear areas had died. In the dishes to which no mold spores had been added, there were no clear areas where bacteria had died.

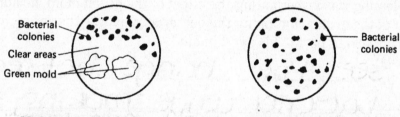

Mold spores added   **No mold spores added**

F. After repeating his work many times, Fleming became convinced that the green mold was in some way responsible for either killing the bacteria or preventing them from reproducing.

G. Today, as a result of the work of Fleming and other scientists, the chemical produced by the green mold has been purified and made into a medicine, which is used throughout the world. The medicine is called *penicillin*. For his work, Fleming was awarded a Nobel prize in 1945.

Analyzing the Experiment

Can you identify each of the steps in the experiment Fleming carried out? In the answer blank after each of the following questions, place the letter (A–G) of the paragraph that describes the step.

1. Which statement describes Fleming's *conclusion*?   __E__

2. Which statement describes his *observations*?   __B__

3. In which statement is the *problem* stated?   __A__

4. Which statement describes the *application* of Fleming's work?   __F__

5. Which statement describes Fleming's *hypothesis*?   __C__

6. In which statement is the *experiment* described?   __D__

TEST YOUR UNDERSTANDING

1. Fleming and all other scientists repeat their work many times. Why is it important for scientists to repeat their work?

   to test w/ diff. variables + make
   sure their results are consistent

2. When Fleming was investigating the effect of the green mold, he added the mold to only half of the dishes containing the bacteria. Why was the set of dishes *without* mold an important part of his experiment?

   to see the change between
   the bacteria while making sure that
   both dishes were tested under the
   same circumstances

   This step is known as the **control** in an experiment.

3. If bacteria had died in the dishes *without* mold, what conclusion would Fleming have drawn from this experiment?

   that the mold did not kill the
   bacteria

4. Once Fleming had proved that penicillin mold has a harmful effect on bacteria, what new questions or problems remained for Fleming and other scientists to solve?

   That the penicillin could
   be used to kill harmful bacteria
   inside the human body.

GOING FURTHER

5. Fleming's discovery of the power of penicillin mold is sometimes described as a "happy accident" because he did not plan that the mold spores would fall onto his petri dish. What arguments can you offer to support the view that his discovery was *not* an accident?

Fleming doing research + testing the mold's affect on bacteria was not accidental.

6. What are some facts you have read?
   a. What is the name of the scientist?

   Alexander Fleming

   b. In what year did he do this important experiment?

   _1928_

   c. What kind of disease **organisms** was the scientist studying?

   bacteria

   d. What kind of organism seemed to kill the disease organisms?

   mold

   e. As a result of this scientist's investigation of a "happy accident," what medicine was developed?

   penicillin

10

Observing: Comparing

✓ WORD CHECK

liquid	one physical form of a substance in which the particles of the substance are held closer together than they are in the gas form
properties	identifying characteristics of a substance, such as boiling and freezing points

PROBLEM: How can you determine the identity of an unknown liquid?

Making careful observations is an essential part of scientific work. Scientists carefully observe things by using their senses (sight, touch, smell, etc.), by measuring with instruments, and by special testing.

As an example, a scientist who wants to identify a **liquid** in an unlabeled bottle would probably begin by observing the color and odor of the liquid. Then, the scientist might measure the temperature at which the liquid begins to boil or freeze. Also, the scientist might conduct tests to see what happens when the unknown liquid is mixed with other chemicals. All these scientific techniques and skills produce observations, information, and data describing **properties**, or identifying characteristics, of the liquid.

Today, you will identify different liquids that are hard to tell apart because they all look alike. You will use your senses, conduct simple tests, and make observations to identify each liquid. By comparing properties you observe in each unknown liquid with properties of known liquids, you can make a positive identification of all the liquids.

EQUIPMENT

First, make sure you have all the equipment and materials in the following list. Place a check next to each item if it is present at your workstation. If any item is missing, obtain it from your teacher.

____ 3 test tubes   ____ red litmus paper

____ test-tube rack   ____ blue litmus paper

____ marking pencil   ____ forceps

____ 3 dropper bottles, labeled A, B, and C

PROCEDURES

One of the dropper bottles contains an **acid**. Another has pure water, a **neutral** substance. The third contains a **base**, a liquid that you tested and identified in a previous lesson. You will determine which type of liquid is in each bottle. Proceed as follows:

1. Observe the liquid in each bottle. Describe each one.

    Bottle A _____

    Bottle B _____

    Bottle C _____

    From their appearance alone, can you tell one liquid from another? Explain.

    _____

    _____

    _____

2. Your teacher will demonstrate the safe way to check the odor of a chemical. The technique is called *wafting* or fanning. Use this safety technique to investigate the odor of the liquid in each dropper bottle. What are your observations?

    _____

    _____

    _____

3. With the marking pencil, label the test tubes A, B, and C. Then use the dropper in bottle A to add a dropperful of the liquid in the bottle to the test tube marked A. Do the same with the bottles and test tubes labeled B and C. Tightly close the top of each dropper bottle immediately after you have finished using it. Why?

    _____

    _____

4. Add to the diagrams at the top of the next page to show that the test tubes now contain liquid.

5. To help you determine the identity of the liquids in the three test tubes, your teacher will demonstrate how special types of paper called *litmus paper* react when exposed to acids and bases.

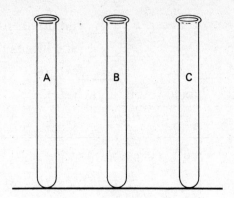

    a.   How does *red* litmus paper change in a base?

_____

    b.   How does *red* litmus paper change in an acid?

_____

    c.   How does *blue* litmus paper change in an acid?

_____

    d.   How does *blue* litmus paper change in a base?

6.   Summarize the way in which litmus paper can be used to determine if a liquid is an acid or a base.

_____

_____

7.   Sometimes it is difficult to remove a piece of litmus paper from a container in which you have tested a substance. What piece of equipment can be helpful in this situation?

_____

_____

8.   Add one piece of red litmus paper and one piece of blue litmus paper to each test tube.

Which test tube contains the acid?

_____

How do you know?

_____

Which test tube contains the base?

_____

How do you know?

_____

Which test tube contains the water?

_____

How do you know?

_____

_____

9. Summarize your observations by completing the table, and then identify the liquids.

Properties of Three Liquids

| Test tube | Observations | | | | Identity of liquid |
	Color	Odor	Effect on red litmus	Effect on blue litmus	
A					
B					
C					

10. Pure water is described as a **neutral** substance. From your tests with litmus paper, what do you think the term *neutral* means?

_____

_____

11. Why can litmus paper be described as a scientific tool, just as glassware and other equipment you use in the laboratory are tools?

_____

_____

_____

TEST YOUR UNDERSTANDING

1. Why is it safer to use wafting to smell a chemical than to smell the contents of a container directly.

_____

_____

2. What should you do if your laboratory partner thinks that a clear, colorless, and odorless liquid in a bottle without a label is water and is about to use the liquid?

    _____

    _____

3. The water in some communities is called *hard* because it contains minerals that make it a weak base. How could you test the water in your home to find out if it is acid, base, or neutral?

    _____

    _____

    _____

4. Tomato plants grow best in acid soil. How could you find out if soil is suitable for growing tomatoes?

    _____

    _____

    _____

GOING FURTHER

5. Take home some pieces of both red and blue litmus paper and test the following substances to find out if they are acids, bases, or neutral substances.

orange juice	_____	water	_____
liquid hand soap	_____	sugar in water	_____
vinegar	_____	alcohol	_____
regular shampoo	_____	salt in water	_____

6. When some shampoos get into the eyes, the eyes sting and water. Why do you think these shampoos irritate the eyes? Base your answer on your test of a regular shampoo with litmus paper.

    _____

    _____

7. Other types of shampoos, such as those used for babies, are mild and do not sting the eyes. How must baby shampoos differ chemically from other shampoos?

    _____

    _____

8. How could you find out if your answer to question 7 is correct? Use the steps in the scientific method, and explain your reasons for carrying out each step.

  a. Problem

  _____

  _____

  b. Hypothesis

  _____

  _____

  c. Experiment

  _____

  _____

  d. Observations

  _____

  _____

  e. Conclusion

  _____

  _____

9. Why should you also test regular shampoo of the type that stings the eyes?

  _____

  _____

  _____

C H A P T E R

11

Reading for Information

☑ WORD CHECK

physical quantity	something that is measurable, such as length, temperature, time, or pressure
standard unit of measure	a unit set up and established by authority, law, or custom as a rule for the measure of a physical quantity

PROBLEM: How do you measure length in science?

Scientific work involves making accurate observations. When scientists perform an experiment or an investigation, they are often required to measure things they are observing. For example, they might want to determine the distance a tiny animal moves over a period of time or how high above the ground a plant grows in different colored lights. In both examples, the **physical quantity** that is being measured is *length*, or the shortest distance between two points—a straight line.

In the work today, you will first be checking to see how well you can judge differences in lengths.

EQUIPMENT

Check to make sure you have a clean sheet of paper and a pencil before you begin. If you do not have these items, obtain them from your teacher.

_____ piece of paper and a pencil

PROCEDURES

Study the figures carefully, and then answer the following questions:

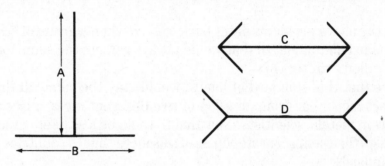

1. Which is longer, line A or line B?

    _____

2. Which is longer, line C or line D?

    _____

Now check to see how accurately you judged the lengths of these lines. Do this by first holding the edge of a piece of paper against line A. Using a pencil, mark off the length of line A on the paper. Then, compare this marked length to the length of line B. Follow the same procedure for line C, and then compare line C with line D.

3. Which is actually longer, line A or line B?

    _____

4. Which is really longer, line C or line D?

    _____

5. After having checked your judgment of length against a piece of marked paper, are you surprised by your results? Explain.

    _____

    _____

    _____

6. Based upon your above observations and comparisons, what conclusion can you arrive at about your ability to judge lengths by eyesight?

    _____

    _____

A more accurate way to measure length, other than to use your eyes or to make marks on a piece of paper, is to use a ruler. Read the paragraphs that follow to gain a better understanding of the history of rulers and the **standard unit of measure** for length, the *meter*.

Being able to measure length accurately in your own life can be almost as important as being able to measure length in science. Imagine, for example, how upset you would be if an article of clothing you wanted shortened was measured to the wrong length.

Many years ago, measuring length was difficult because the standard unit of measure could change from year to year. In fact, units of length were often based upon a part of a king's or an emperor's body—such as the length of his thumb or arm. You can see how easily this might have caused problems, especially when kings of different sizes replaced one another.

A solution to the above problems dates back to 1793 when a group of scientists in France developed a standard unit of length that would always remain the same for everybody. The unit of length is called a **meter (m)**.

To make sure that this standard of length would stay the same all the time, scientists made a very exact meter bar from an *alloy* of two different metals. Because this bar was made from different metals, scientists knew that it would be less likely to suffer any changes in length. Scientists then sent accurate copies of this meter bar to countries all over the world to be used as standards.

The meter remains the standard unit of length in most countries in the world and in science. Today, however, there are better ways to determine the length of a meter. We no longer have to make copies of the meter bar that is still in France.

In your course of study, you will measure objects and materials for many of their physical characteristics and properties, such as volume, mass, temperature, and density. In every instance, you will be using standard units of measure that make up the *Metric System*.

TEST YOUR UNDERSTANDING

1. Without using a ruler, a friend judged a shelf to be "three meters above the floor." Why could you correct your friend by adding "about" three meters above the floor?

    _____

    _____

    _____

2. In the days when parts of a king's body were used as standard units of length, who would have been more pleased if a newly crowned king had short arms—a merchant who was selling rugs or a customer who was buying them? Why?

    _____

    _____

    _____

3. Why is the meter described as the "standard" unit of length? ____ (Write the letter of the correct choice.)

    a. The meter can be divided into many smaller lengths.
    b. The meter is used in many countries and in science.
    c. The meter was developed about two hundred years ago.
    d. The meter can be used to measure very long lengths.

GOING FURTHER

4. Both the standard meter bar in Franch and a quarter or dime of today are *alloys*. How must they be alike?

    _____

    _____

5. How did countries throughout the world make sure that the length of a meter in one country was the same length as a meter in another country?

    _____

    _____

    _____

C H A P T E R
12

Reading a Scale

WORD CHECK

scale	a series of lines or points that marks off known spaces, or distances, on a measuring instrument, such as the ruler

PROBLEM: How do you read the scale on a metric ruler?

As you learned in another worksheet, trying to judge lengths with your eyes can lead to mistakes. A more accurate way to measure length is with a metric ruler. In your work today, you will be studying a ruler closely to discover the types of information it provides.

EQUIPMENT

Check to see if you have a metric ruler. If you do not, obtain a metric ruler from your teacher.

_____ metric ruler

PROCEDURES

The first thing to observe about the ruler is that, like other measuring instruments, it has a **scale**. The scale is composed of a series of evenly spaced lines that mark off the ruler into equal parts. On the scale in the figure, the longest numbered lines mark off centimeters (cm). That is, the space from one numbered line to the next numbered line is equal to a length of 1 centimeter. The first part of this word, **centi-**, means one-hundredth. The second part of this word, **-meter**, you have learned is the standard unit of length in the metric system of measurement. A **centimeter** is one one-hundredth of a meter. A meter is, therefore, made up of 100 centimeters. The figure is marked off into 15 cm. Check the length of the scale in the figure with the ruler.

1. In the figure, how many small spaces are marked off within 1 centimeter, for example, between the lines numbered 7 and 8?

_____

These smaller spaces are called millimeters (mm).

2. How many millimeters are there in a meter?

_____

3. What do you think the prefix milli- means?

_____

Now you'll see how to read and use the scale on the ruler.

The metric system of measurement is a decimal system, which means that it is based on units of 10. Our money system is also a decimal system. The basic unit, the dollar, is made up of 10 dimes, each of which is made up of 10 cents. Portions of a dollar are written as decimals, as $7.10 or $0.50. Metric units are written in the same way.

When the length of an object is less than 10 on the scale, you indicate the measurement by writing a decimal point and a zero after the measurement. In the figure, the arrow indicates a length of 7.0 cm, not 7 cm.

When the length of an object reaches one of the smaller lines on the scale, the fraction of the unit is written as a decimal. The arrow indicates a length of 5.8 cm.

When the length of an object falls between two lines on the scale, the measurement can be estimated. In this enlarged sample, the arrow indicates a measurement of 4.3 – 4.4 cm (or 4.35 cm).

The scale on some rulers does not begin at the edge of the ruler. Such rulers can give accurate measurements even if the edge of the ruler chips off. Be careful that your readings always start at the beginning of the scale, not at the beginning of the ruler.

When the length of an object is less than 1.0 cm, the measurement must still be written with two numbers, such as 0.7 cm, not as .7 cm.

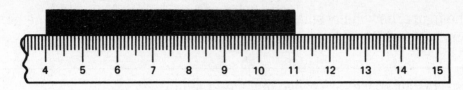

4. The ruler above is broken, but part of the scale that remains can still be used to measure the length of an object. The piece of tape in the figure stretches from the 4.0 cm mark to the 11.0 cm mark. By subtracting the two numbers, you know the length of the object is ____.____ cm.

TEST YOUR UNDERSTANDING

What length is indicated by the arrow in each of the following figures?

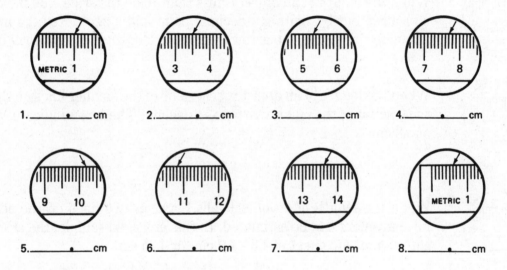

1._____.____ cm   2._____.____ cm   3._____.____ cm   4._____.____ cm

5._____.____ cm   6._____.____ cm   7._____.____ cm   8._____.____ cm

9. How long is the earthworm? _____.____ cm.

10. Describe how you arrived at your answer to question 1.

_____

_____

Diagrammed above are sections of two different types of rulers. Both could be used to measure lengths but their scales are different.

11. The scale on ruler A is the same as those illustrated earlier on this worksheet.
    • On ruler A, how many spaces are there between each of the numbered lines?

    _____

    • How does this differ on ruler B?

    _____

    _____

GOING FUTHER

12. To use a scale such as the one shown on ruler B, follow these procedures:
    • Write down the number that appears after the object's length   7.0 cm
    • Then subtract from this, the number that appears just before   - 6.0 cm
    • The difference is   1.0 cm
    • Now divide this difference by the number of *spaces* (5) between   0.2 cm
      the two numbers.   5) 1.0 cm
    • This resulting number (0.2) is to be added to the number just before
      the object's end (6.0) enough times to reach the line at the end of the
      object. This is the total length of the object.

$$6.0 + 0.2 + 0.2 + 0.2 = 6.6 \text{ cm}$$

C H A P T E R
13

Measuring Length

✔ WORD CHECK

| dimension | an object's height, length, or width |

PROBLEM: How do you measure length with a metric ruler?

In another lesson, you learned how to read the scale of a metric ruler. Today, you will use that knowledge to make measurements of real objects.

EQUIPMENT

Check to see if you have a metric ruler. If you do not, obtain a metric ruler from your teacher.

_____ metric ruler

PROCEDURES

You and your partner will make two sets of measurements, one set at your desk and the other set at different locations in the classroom. Each object to be measured in the classroom is identified by a label that has a number and tells you what **dimension** to measure. The numbers on the labels match the numbers of the blanks to be filled in under the heading *Measurements of Classroom Objects*. Write the name of the object in the blank provided for it (item 6 has been filled in as an example), and record the measurement specified by the label. For example, you should proceed as follows when measuring the length of an object:

- If an object's length is longer than your ruler, make a light pencil mark on the surface of the object to show the end of the ruler's scale. Then, place the beginning of the scale at the mark and read the measurement.

- If the ruler still does not reach the end of the object's length, mark the end of the scale again. Repeat the process until you have measured the entire length of the object. Then, add the individual measurements to obtain the total length.

- Remember, if the length of an object is less than 10 cm, the measurement should be recorded as two numbers (for example, 0.8 cm, 4.7 cm, 6.0 cm).
- If the length of an object falls between two lines on the scale of the ruler estimate the measurements (for example, 14.9–15.0 cm).
- If too many students are waiting to measure a particular object, move on to another one or return to your seat and work on the section *Measurements to Be Made at Your Desk*. You can later return to make measurements of objects you had to skip over.

Measurements to Be Made at Your Desk:

1. How long is this box? (L)   _____.____ cm

2. How wide is this box? (W)   _____.____ cm

3. How high is this box? (H)   _____.____ cm

4. How wide is your book?   _____.____ cm

5. How long is your book?   _____.____ cm

Measurements of Classroom Objects:

6. Width of the classroom door _____ _____.____ cm

7. _____ _____.____ cm

8. _____ _____.____ cm

9. _____ _____.____ cm

10. _____ _____.____ cm

11. _____ _____.____ cm

12. _____ _____.____ cm

13. _____ _____.____ cm

14. _____ _____.____ cm

15. _____ _____.____ cm

TEST YOUR UNDERSTANDING

1. Why is it an advantage to make measurements as part of a team, instead of working alone?

_____

_____

_____

2. What should you do if you and your partner obtain different measurements for the same object? Why?

_____

_____

GOING FURTHER

Read the following description of a problem a group of students met when making measurements and the steps they took to solve the problem. Then, match the letter in front of the steps to the scientific process the students used.

Students in a science class were asked to measure different objects in their classroom, using plastic measuring tapes. They worked very carefully in teams and checked with one another to make certain that the measurements of team members agreed. But when all teams reported their measurements, the students were surprised to find that the results of different teams differed by as much as 2.0 cm. They decided to look for an explanation.

A. The students wondered if the differences in results were due to differences in the plastic measuring tapes.

B. The students guessed that constant stretching of the tapes when students in other classes used them could have changed the length of the tapes.

C. To find out if their guess was correct, the students held the tapes side by side to compare the length of the tapes.

D. They also checked the length of the tapes against a wooden meterstick.

E. The students saw that all the tapes were slightly different in length and that none was the same length as the wooden meterstick.

F. They repeated their work a number of times with other tapes that also had been used previously. Their results convinced them that the plastic measuring tapes had stretched and, therefore, no longer gave accurate measurements.

G. Now when the students in this class make measurements, they either use wooden metersticks, or they check the length of their plastic tapes against metersticks before they begin.

Which one of the steps, A through G, describes the students'

3. conclusion?   ____   6. application?   ____   8. experiment?   ____

4. problem?   ____   7. observations?   ____   9. hypothesis?   ____

5. control?   ____

14

Measuring the Mass of a Solid Object

WORD CHECK

gravity	includes the force of attraction between the earth or moon and objects on their surface

PROBLEM: How do you use a balance?

Measuring **weight** is an important part of daily life. Dieters, for example, want to find out if they have gained or lost weight. Meat in a supermarket is weighed before it is purchased. In some states, a car's registration fee is based upon the weight of the vehicle.

Being able to weigh something is also important in science. However, in scientific terms what is actually being measured is **mass** rather than weight. The difference between mass and weight is described as follows:

When we see videos of astronauts moving about on the surface of the moon, we are always amazed at how high they are able to jump with little effort. It's almost as if they had lost weight. That is exactly what has happened.

The moon is much smaller than Earth and, therefore, has less pull of **gravity**. Since weight depends upon gravity, things weigh less on the moon than on Earth. However, an astronaut who weighs less on the moon is still made of the same amount of material or mass. The mass of an astronaut does not change whether the astronaut is on Earth or on the moon.

In summary, weight depends upon gravity and can change as gravity changes. But the mass of an object remains the same no matter how gravity changes. For this reason, scientists measure mass rather than weight.

EQUIPMENT

Check off the items to make sure you have the following equipment and materials. If any item is missing, obtain it from your teacher.

_____ triple-beam balance   _____ penny, nickel, dime

PROCEDURES

The standard unit of mass in the metric system is the **kilogram (kg)**, a thousand grams. A gram is one-thousandth of a kilogram. The metric symbol for the gram is **g**. Different types of instruments are used to measure mass, but the one being described is a *triple-beam balance*.

SAFETY: Before starting the lesson, your teacher will instruct you in the correct procedures for carrying and using the balance safely.

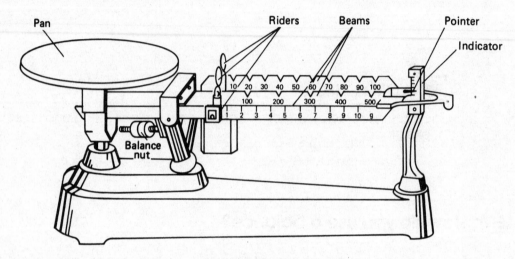

Triple-beam balance

1. Place the balance before you in the same position, as the one in the figure. The pan should be on the left. Compare the figure of the triple-beam balance to the balance you have before you. Locate all the labeled parts. Why do you think this instrument is described as a *triple*-beam balance?

    _____

    _____

2. If only the beam nearest you is used, what is the greatest mass of an object that you could find?

    _____ g

    If only the beam farthest away from you is used, what is the greatest mass of an object that you could find?

    _____ g

    What is the greatest mass of an object that you could find if only the middle beam is used?

    _____ g

    What is the total mass an object could have if all three beams are used?

    _____ g

3. When preparing the balance for use, you should always do the following:

    • Make sure the riders are set at zero.

    • Fit each rider into the notch on its beam.

    • The pointer should match up exactly with the indicator. If the pointer does not, you must use the balance nut or wheel under the pan end of the beams.

    How does the pointer move when the balance nut is tightened?

    _____

    _____

    Why must you always check this adjustment before you start using the balance?

    _____

    _____

4. Reading the scales on the beams of the balance is the same as reading the scale on a metric ruler. By using previous measurement techniques, determine the masses of objects A, B, and C when the riders are positioned on the beams as shown in the figures.

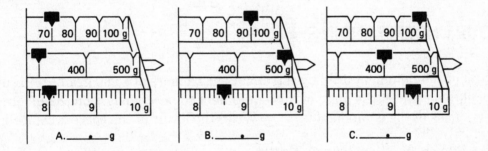

5. You are now going to determine the mass of a penny, a nickel, and a dime. What should you do if your penny has a mass that is more than a mass that can be measured, using just the first beam?

    _____

    _____

    _____

6. Place the coin in the center of the pan and then record the observations of your measurements:

    mass of your penny   _____ . _____ g.

    mass of your nickel   _____ . _____ g.

    mass of your dime   _____ . _____ g.

    Find the mass of your penny, nickel, and dime all together.   _____ . _____ g.

Why do this if you have already found their masses separately?

_____

_____

_____

_____

What should you do if the total mass of the three coins together does not equal the sum of the individual masses of the three coins? Why?

_____

_____

_____

_____

TEST YOUR UNDERSTANDING

1. A piece of equipment used by astronauts has a mass of 1,000 grams on Earth. What would you expect the mass of this equipment to be when it is on the moon? The mass of the equipment is ____ (Write the letter of the correct choice.)

    a. less on the moon than on Earth.

    b. the same on the moon as on Earth.

    c. more on the moon than on Earth.

2. What would you expect the weight of this equipment to be when it is on the moon? The weight of the equipment is _____ (Write the letter of the correct choice.)

    a. less than 1,000 grams.

    b. exactly 1,000 grams.

    c. more than 1,000 grams.

3. Briefly describe why you answered questions 1 and 2 as you did.

    _____

    _____

    _____

GOING FURTHER

4. Assume that you and another team member determined the mass of two different pennies equally carefully. Why might the actual mass of his or her penny not be the same as yours?

_____

_____

5. Suppose you were standing on the surface of a planet that was much larger than Earth. How would the height to which you can easily jump on the planet compare to the height to which you can easily jump on Earth? _____ (Write the letter of the correct choice.)

   a. You can jump higher on the planet.
   b. You can jump to the same height as on Earth.
   c. You can jump higher on Earth.

6. Explain your answer to question 5.

_____

_____

_____

_____

_____

C H A P T E R

15

Observing the Surface Properties of Water

WORD CHECK

particle	a very small portion or amount of a substance

PROBLEM: Why is the surface of water unusual?

Water is such a common substance that we tend to take it for granted unless a serious problem like a flood or water shortage occurs. But in science we can learn a lot about water.

In the worksheets that follow, you will be working with water in a number of different ways. Your activities will include pouring water, heating and cooling it, displacing it, measuring its volume, and determining its mass and its density.

As familiar as water is to you in your daily life, you have probably let the observation of the way in which water **particles** affect each other go unnoticed. Today you will be exploring this particular characteristic of water.

EQUIPMENT

Check off the items to make sure you have the following equipment and materials. If any item is missing, obtain it from your teacher.

_____ dropper bottle of water   _____ paper clip

_____ aluminum foil (small piece)   _____ petroleum jelly

_____ pencil   _____ length (15 cm) of sewing thread

_____ paper or plastic cup   _____ dropper bottle of liquid soap

_____ shallow pan

PROCEDURES

1. Use a medicine dropper to add a single drop of water to a small, flat piece of aluminum foil. Observe the drop from the side, and then draw what it looks like. This effect is called *beading*.

_____
Drop of water

2.   How must water particles on the surface of the drop be reacting to each other to produce this beading effect?

   _____

   _____

3.   Add a second drop of water so that it is close to the first drop but not touching it. With the point of a pencil or pen, move one of the drops slowly toward the other until both drops just barely touch. What happens as the drops touch?

   _____

   _____

   _____

   Why do you think this took place?

   _____

   _____

   _____

4.   Place a paper or plastic cup in a shallow pan and fill the cup with water until the level of the water is exactly even with the rim of the cup. Using a medicine dropper, slowly add more water until the water finally begins to overflow. Draw the surface of the water just before it began to overflow. Why do you think the water particles were able to stay together for so long before spilling over the edge of the cup?

   _____

   _____

5.   Tie a length of sewing thread (15 cm) to one end of a paper clip. Hold the paper clip (to which a little petroleum jelly has been added) lengthwise with your thumb and index finger, and then carefully and gently lay the paper clip flat on the surface of the water. What happens?

   _____

   _____

   _____

   If the paper clip sinks and you have to repeat the procedure, use the thread to retrieve the paper clip. Why do you think the paper clip was unable to break through the surface particles of the water?

   _____

   _____

6. The skinlike property of the water's surface, produced by the sticking together of water particles, is called **surface tension**. Add one or two drops of liquid soap to the cup of water while the paper clip is still floating on the surface. What happens?

_____

_____

7. How must the soap have changed the way in which the water's surface particles stick to one another?

_____

_____

_____

TEST YOUR UNDERSTANDING

For questions 1 and 2, include the term *surface tension* in your answer.

Water
Boatman

1. Some insects have the ability to walk, without sinking, on the top of pond water. What must be true about these insects if this is possible?

_____

_____

_____

2. Commercial laundries sometimes release their soapy waste water into nearby ponds and streams. How could a study of water-walking insects just described in question 1 help to detect soap pollution in our waters?

_____

_____

_____

3. Give an example of the beading effect of water observed around your home.

_____

_____

_____

4. On a rainy day, how can you identify a car that has been recently waxed and polished?

_____

_____

Why do you think this happens?

_____

_____

_____

GOING FURTHER

5. Why should it be an advantage for plants to have leaves with shiny, waxy upper surfaces when there is a heavy rain?

_____

_____

6. In the work today, you observed how water particles are strongly attracted to one another and how they stick together to form a skinlike covering on the surface of the water. What hypothesis would you form to predict what happens when water particles come into contact with the surface of a substance such as glass?

_____

_____

_____

To find the answer to your hypothesis, you will be studying a **meniscus** in your next worksheet.

C H A P T E R
16

Measuring the Volume of a Liquid

☑ WORD CHECK

volume	the amount of space that a substance occupies

PROBLEM: How do you use a graduated cylinder to measure the volume of a liquid?

You may have read a food label that stated, "This box of cereal is sold by weight and not by volume." The term **volume** is being used to describe the amount of space taken up by the cereal. The message on the label is intended to reassure customers that they are not being cheated when they see that the volume of the cereal does not completely fill the package.

The word *volume* is used not only to describe solids, such as cereal, but also gases (the amount of air in a balloon), and liquids (the amount of soda in a bottle).

In your work today, you will learn how to use a *graduated cylinder* to measure the volume of a liquid.

EQUIPMENT

Check off the items to make sure you have the following equipment and material. If any item is missing, obtain it from your teacher.

_____✓_ 10-mL graduated cylinder   _____✓_ dropper bottle of water

PROCEDURES

As you examine the graduated cylinder, note the series of lines or markings on the side. These markings make up the measurement scale and are called graduations. Therefore, the instrument is known as a **graduated cylinder** (see next page).

1. Next, note the *lip* at the top of the cylinder. Why is it an advantage for the cylinder to be shaped with a lip at the top?

easier to pour liquids

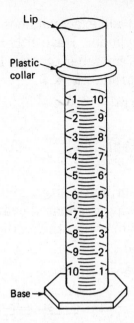

Lip

Plastic collar

Base →

2. A *plastic collar* may be wrapped around the cylinder. Why is this collar a useful safety feature?

    <u>to avoid chemicals dripping on you.</u>

3. The base of some graduated cylinders are circular in shape. Others, as shown in the figure, have bases with six sides. Why is the *six-sided base* a useful safety feature?

    <u>stability is increased</u>

4. The standard unit of liquid volume in the metric system is the **liter (L)**. Just as the standard unit of length, the meter, was divided into smaller units called millimeters, so also is the liter divided into smaller units called **milliliters (mL)**. How many milliliters can your graduated cylinder measure at one time?

    <u>100mL</u>

5. Use a medicine dropper to fill your graduated cylinder about halfway with water. Then, from the side, look at the upper surface of the water, as shown in the figure. What do you notice about the water's surface that is unusual?

    <u>it appears to dipdown</u>

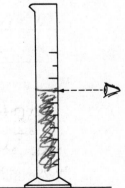

6. Draw the water's surface in the figure. It is called a **meniscus**. When reading the scale on a graduated cylinder, you must always use the *bottom* of the curve of the meniscus.

7. Use the medicine dropper to add exact volumes of water (3.0 mL, 5.0 mL, and 7.0 mL) to your graduated cylinder. Have either your partner or teacher check off your success on this sheet. Remember that the bottom of the meniscus must sit exactly on top of the line or graduation.

3.0 mL ___✓___   5.0 mL ___✓___   7.0 mL ___✓___

TEST YOUR UNDERSTANDING

Graduated cylinders are made in many sizes, and the scale on each size may differ considerably. For each of the following graduated cylinders, identify the volume of liquid shown in the figure. It may help you to review the method you used to determine the values of lines and spaces on the metric scale.

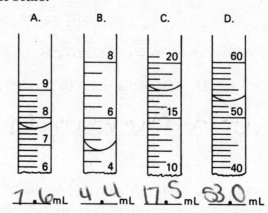

A. 7.6 mL   B. 4.4 mL   C. 17.5 mL   D. 53.0 mL

GOING FURTHER

1. The graduated cylinder you used today has a series of numbers that started at the base. The numbers increased in value as you read them upward and toward the top of the cylinder (for example, 1 mL up to 10 mL). You used these numbers to determine the volume of the water added to the cylinder. Many cylinders have another set of numbers that increase in value as you read them downward from top to base (for example, 1 mL down to 10 mL). (Check yours ___ ___✓___) When would this second set of numbers be useful?
   yes   no

   when recoading diffent amounts of a liquid

2. What are eight examples of liquid products sold by volume?

   soda, water, meds, hair products, lotions, cleansers, milk,

3. The prefix **milli-** means 1/1,000. How many times would you have to fill and empty a 10-mL graduated cylinder in order to fill a 1-liter soda bottle?

   ___100___ times

17

Measuring the Mass of a Liquid

✓ **WORD CHECK**

error	the difference between an observed or calculated value and the true value

PROBLEM: How do you determine the mass of a liquid?

Using the triple-beam balance to measure the mass of a solid is a skill you developed in previous lessons. However, using this instrument to measure the mass of a liquid presents an obvious problem—that of getting the liquid to stay on the pan.

The answer is similar to the way you determine your dog's mass when it won't stay put on your bathroom scale. First, you determine your mass alone. Then, you determine your mass again while holding your dog in your arms. The difference between these two measurements is the mass of your pet.

In your work today, you will first determine the mass of a graduated cylinder by itself. You will then use the graduated cylinder to hold a liquid while you determine their combined mass. The difference between the two measurements is the mass of the volume of liquid.

EQUIPMENT

Check off the items to make sure you have the following equipment and material. If any item is missing, obtain it from your teacher.

_✓_ triple-beam balance _✓_ 10-mL graduated cylinder
_✓_ dropper bottle of water

PROCEDURES

1. What is the mass of the graduated cylinder before any water has been added to it?

   25.99 g
   26

2. Add water to the graduated cylinder until it is about one-third full. Use the medicine dropper to do this. What is the volume of the water in the graduated cylinder?

    30 . _____ mL

3. Find the mass of the graduated cylinder again, but this time with the water in it.

    35 . _____ g

4. Subtract the two measurements of mass.

    mass of the cylinder with water in it   35 30 . _____ g

    – mass of the cylinder by itself   26 –26 . _____ g

    mass of this volume of water   10 9 . _____ g

5. Based upon the information you have just collected, what hypothesis can you form about a volume of water and the mass of that volume of water?

    10 mL of water weighs roughly 10 gs.

6. To test your hypothesis, repeat the procedure with different volumes of water. You should have a total of three additional trials. Complete the table with your results.

	Volume of water		
	8 . _____ mL	6 . _____ mL	7 . _____ mL
Mass of graduated cylinder with water	26 . _____ g	26 . _____ g	26 . _____ g
– Mass of cylinder itself	34 . _____ g	–32 . _____ g	–33 . _____ g
Mass of water	8 . _____ g	6 . _____ g	7 . _____ g

TEST YOUR UNDERSTANDING

1. Why should you measure the mass of the graduated cylinder *before* a liquid is added to it rather than after a liquid has been poured out of it?

    there may be some drops of water left-over

2. Why should you use a medicine dropper to add the water to the graduated cylinder rather than pour the water from a beaker into the graduated cylinder?

   for a more precise number a amount

3. What precaution did you have to take when reading the volume of the liquid at its surface?

   to measure at the meniscus

4. A graduated cylinder has a mass of 28.2 g with 2.1 mL of water in it. What is the mass of the cylinder alone without any water in it?

   26.1 g.

5. What is the mass of 1.0 mL of water?

   1.0 g.

GOING FURTHER

6. Suppose your laboratory work revealed that 5.0 mL of water has a mass of 4.8 g. What **error** might you have made in the use of your triple beam balance?

   round up?

   if you may not have reset your balance

7. Referring to question 6, what error might you have made in determining the mass of the graduated cylinder?

   you may have measured your scale w water already in it

8. What should you do to arrive at the correct answer in question 6?

   reset your scale

18

Determining the Volume of a Regularly Shaped Solid

73

✓ WORD CHECK

square	a rectangle with all four sides equal
cube	a regular solid with six equal square sides
cubic centimeter	the volume of a cube whose edge is 1 centimeter

PROBLEM: How do you determine the volume of a cube or block?

What do the shapes of the following objects have in common?

     a box of cereal   a brick

     a bar of butter   a lump of sugar

In answering this question, you probably responded that all of these objects have regular shapes with six flat sides.

Since these solids, like liquids, take up space, they also have volumes. Unlike the liquids, however, the volumes of these solids cannot be measured using graduated cylinders because of the solids' sizes and shapes. Therefore, in your work today, you will use a different technique and a metric ruler to determine the volume of a regularly shaped solid.

EQUIPMENT

Check to make sure you have a metric ruler. If you do not, obtain a metric ruler from your teacher.

    ✓   metric ruler

PROCEDURES

Use a metric ruler to measure the diagrams that follow. Measure all diagrams to the nearest whole number (for example, 2.0 cm instead of 1.9 cm or 2.1 cm).

KHDUDCM

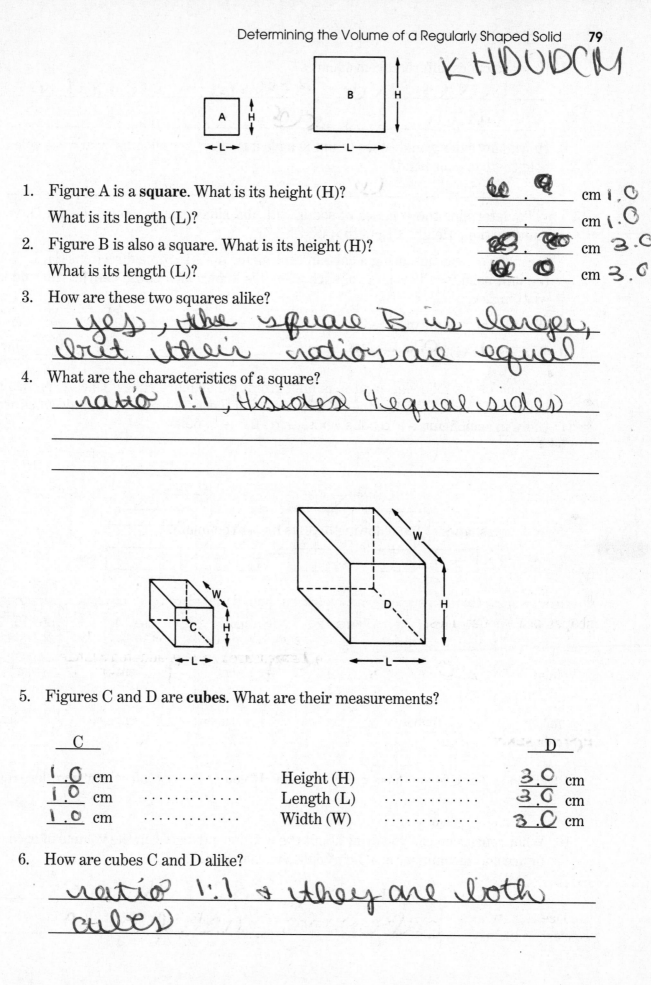

1. Figure A is a **square**. What is its height (H)? _____6_._4_ cm 1.0

   What is its length (L)? _____._____ cm 1.0

2. Figure B is also a square. What is its height (H)? __23_._80_ cm 3.0

   What is its length (L)? ___02_._0_ cm 3.0

3. How are these two squares alike?

   yes, the square B is larger,
   but their ratios are equal

4. What are the characteristics of a square?

   ratio 1:1, 4 sides (4 equal sides)
   _____
   _____

5. Figures C and D are **cubes**. What are their measurements?

C		D
1.0 cm · · · · · Height (H) · · · · · 3.0 cm
1.0 cm · · · · · Length (L) · · · · · 3.0 cm
1.0 cm · · · · · Width (W) · · · · · 3.0 cm

6. How are cubes C and D alike?

   ratio 1:1 & they are both
   cubes

7. How are cubes different from squares?

   cubes are 3 demensional +
   have 6 sides

8. How many sides would cubes C and D have if they were real and if you were able to hold them in your hand?

   _6_

   To determine the volumes of solids with flat sides, such as cubes C and D, you multiply their Height × Length × Width.

   Since you are measuring a cube in centimeter units in *three* directions (H, L, W), the unit used for the volume of such a solid is known as a **cubic centimeter** and is written as **cm**^{3}.

9. What is the volume of cube C?   _1.0_ cm^3

10. What is the volume of cube D?   _9.0_ cm^3

   Each of the following figures is built from two or more cubes. Each cube has a volume of 1.0 cm^3. What is the volume of each of these figures? How many cubes are there in each figure?

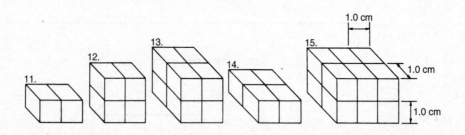

	Figure 11	Figure 12	Figure 13	Figure 14	Figure 15
Volume	2.0 cm^3	8. cm^3	4. cm^3	4. cm^3	12. cm^3
Number of cubes	2 cubes	8 cubes	4 cubes	4 cubes	12 cubes

16. What conclusion can you draw about the relationship between the volume of each figure and the number of 1.0-cm^3 cubes in each figure?

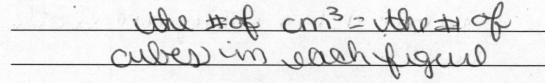

   the # of cm^3 = the # of
   cubes in each figure

TEST YOUR UNDERSTANDING

1. Use a metric ruler to measure the height, length, and width of the following blocks. What is the volume of each of the blocks below?

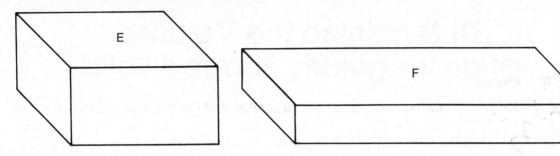

2. How many 1.0-cm³ cubes would you need to construct each one?

	Length	Height	Width	Volume	Number of cubes
Block E	2. cm	2. cm	4. cm	16 cm	16 cubes
Block F	2. cm	1. cm	8. cm	16 cm	16 cubes

3. Blocks E and F have very different shapes but the same volumes. How is this possible?

    <u>their total volume is equal</u>
    <u>in Block F there is a greater w.</u>
    <u>but smaller h + in Block E there is</u>
    <u>a greater h</u>

GOING FURTHER

4. What would be the volume of a block that was built from 27 of the 1.0-cm³ cubes?

    27 . 0 cm³

5. If the block described in question 4 were a cube, what would be its height?

    3 . 0 cm

    Its length?

    3 . 0 cm

    Its width?

    3 . 0 cm

6. A block is 4.0 cm in height; 2.0 cm in length, and 8.0 cm in width. What is its volume?

    32 . 0 cm³

    How many 1.0-cm³ cubes would it take to build it?

    32 . 0

C H A P T E R

19

Determining the Volume
of an Irregularly Shaped Solid

3/3

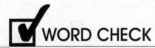 WORD CHECK

| Archimedes | a Greek mathematician, scientist, and inventor, who lived during the third century B.C. |

PROBLEM: How do you determine the volume of a solid that has uneven sides?

In a previous worksheet, you learned how to determine the volume of solids that were cubes or blocklike in shape. These solid objects had regular flat sides, which were easy to measure with a metic ruler. However, solids such as rocks and nails cannot be measured in the same way because of their uneven shapes. In your work today, you will be using a technique called **displacement** to determine the volume of such solids.

Displacement was discovered by the scientist **Archimedes**. The story of the discovery of this technique concerns what supposedly happened to Archimedes while he was taking a bath. Archimedes noticed that the further down he lowered himself into his bathtub water, the higher the level of the water rose. From his observations, he developed the hypothesis that the amount of water pushed out of the way, or displaced, to make room for an object is the same as the volume of the object.

Using a graduated cylinder and Archimedes' displacement technique, you will measure the volume of solids with uneven (irregular) shapes.

EQUIPMENT

Check off the items to make sure you have the following equipment and materials. If any item is missing, obtain it from your teacher.

☑ one nail and three other solids with uneven shapes

☑ 10-mL graduated cylinder

PROCEDURES

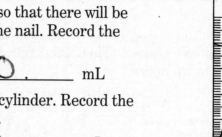

1. Add water to the graduated cylinder so that there will be enough water to cover the height of the nail. Record the volume of the water.

   _10._____ mL

2. Next, add the nail to the water in the cylinder. Record the volume of the water with the nail in it.

   _16._____ mL

3. How has the volume in the cylinder changed?

   ___↑ 16 mL_____

4. Why has the change in question 3 taken place?

   ___the object has a volume of___
   ___6mL_____

5. How can you determine the exact volume of the nail you just added to the graduated cylinder?

   ___subtract the new water___
   ___height by the origional one___
   _____

6. Repeat this process with a number of other solids that have uneven shapes, and determine the volume of each.

	Solid A	Solid B	Solid C	Solid D
Volume of the water with the solid in it	16 mL	17 mL	86 mL	87 mL
– Volume of the water alone	– 10 mL	– 16 mL	– 85 mL	– 86 mL
Volume of the solid	6 mL	1 mL	1 mL	1 mL

7. Why should your partner repeat each of the measurements you have just made?

   ___to dable check the accuracy___
   _____
   _____

TEST YOUR UNDERSTANDING

Sometimes the unevenly shaped solid whose volume you want to determine doesn't fit into a graduated cylinder. Then the process and the equipment have to be somewhat different, as shown below.

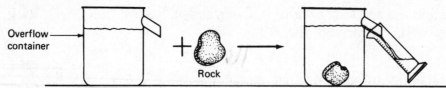

1. What is the purpose of the tube coming out of the side of the overflow can?

   to find the exact volume of the rock

2. Why is the water collected in a graduated cylinder as the water starts to overflow?

   to measure the amount of water the rock displaces

3. Why is this process still an example of measuring the volume of a solid by displacement?

   the amount of water displaced is collected in the graduated cylinder

GOING FURTHER

4. Complete this sentence: Displacement is a process by which

   the initial amount of water is subtracted by the amount after an object is added and the result is the object volume

5. Complete this sentence: The displacement technique is important because it enables you to

   find the volume of objects that aren't square

UNIT 4
DETERMINING DENSITY

C H A P T E R

20

Determining the Density of a Liquid

☑ WORD CHECK

identifying property	a unique or distinguishing characteristic that serves in identifying an unknown substance

PROBLEM: How do you determine the density of a liquid?

There is an old saying that "blood is thicker than water." What this expression means is that members of a family usually feel closer to one another than they do to strangers. Actually blood is not "thicker" than water. Blood has a greater *density* than water.

In today's work, you will discover the meaning of the term *density*. Your experiments will also help you to understand how density can be used as an **identifying property** of a substance—similar to the way fingerprints are used to identify an individual.

You learned in a previous worksheet how to use a triple-beam balance and a graduated cylinder when determining the mass of various volumes of water. Today, you will be using the same technique to determine the mass and then the density of specific volumes of water.

EQUIPMENT

Check off the items to make sure you have the following equipment and materials. If any item is missing, obtain it from your teacher.

___ triple-beam balance   ___ graduated cylinder
___ dropper bottle of water

PROCEDURES

- Find the mass of 1.0 mL, 2.0 mL, 5.0 mL, and 8.0 mL of water. Write the numbers in the table below. Be as exact as you can when measuring the volumes and masses of water.

	Volume of Water			
	1.0 mL	2.0 mL	5.0 mL	8.0 mL
Mass of graduated cylinder with water	27.05 g	29.05 g	31.05 g	34.05 g
–Mass of cylinder itself	–26.05 g	–26.05 g	–26.05 g	26.05 g
Mass of water	1.0 g	2.0 g	5.0 g	8.0 g

- The density of a substance is determined by comparing the two properties you have just been measuring—mass and volume. The following formula describes the density of a substance:

$$Density = \frac{mass}{volume} \text{ or } Density = volume \overline{)mass}$$

- The formula for density shows a relationship between the mass and volume of a substance. Therefore, the units of measure for both mass and volume must be included when describing the density of a substance.
- The unit for mass is the **gram** (g).
- The unit for volume is a little more complicated. Volume is represented by the size of a cube that would be filled with exactly 1.0 mL of water. Such a cube can be found to measure 1.0 cm in all *three* directions (height, length, and width), as shown in the figure.

- Therefore, the unit of measure for volume when used in determining density is the **cubic centimeter,** or **cm**^{3}.

- In summary, the unit of measure for density (mass/volume) is expressed as **g/cm**^{3}.

TEST YOUR UNDERSTANDING

1. What is the density of the water for each of the volumes and masses in today's experiment?

   1.0 mL) $\dfrac{1. \quad \text{g/cm}^3}{1. \quad \text{g}}$ 5.0 mL) $\dfrac{5. \quad \text{g/cm}^3}{5. \quad \text{g}}$

   2.0 mL) $\dfrac{2. \quad \text{g/cm}^3}{2. \quad \text{g}}$ 8.0 mL) $\dfrac{8. \quad \text{g/cm}^3}{8. \quad \text{g}}$

   (If your measurements were not exact, round them off to become whole numbers. For example, 4.9 becomes 5.0, and 8.1 becomes 8.0).

2. Complete the following sentence:
   The density of water is always ____1____ because its __volume__ increases as its __mass__ increases.

GOING FURTHER

Mercury is a very unusual substance. It is a metal, but it is the only metal that is in liquid form at room temperature. Mercury is also a very dangerous chemical. That's why the work to determine its density is being described but not demonstrated. Use the following information to determine its density.

The mass of a graduated cylinder containing 4.0 mL of mercury is 84.6 g. The mass of the graduated cylinder without the mercury is 30.2 g.

3. What is the mass of the mercury in the cylinder? ___64 . 4___ g

4. What is the density of the mercury? ___54 . 4___ g/cm^3

5. The figures on page 88 compare the surfaces of water and mercury inside graduated cylinders. What is unusual?
   __mercury bubbles upwards but water dips down__

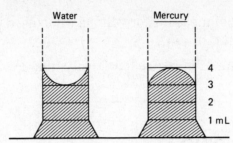

6. Based upon the work you did today, what hypothesis might you form to explain the shape of the mercury meniscus?

mercury is denser than water

21

Determining and Comparing Liquid Densities

☑ WORD CHECK

phenomenon	a fact or event of scientific interest
unique	being the only one

PROBLEM: How do the densities of different liquids compare with one another?

When a day is moving as "slowly as molasses," the day seems to be taking forever to pass. One of the reasons that molasses moves slowly is that it is a dense liquid.

Today, you will determine the specific densities of three liquids. You will then compare them with water. Because its density is 1.0 g/cm^3, water is used as a standard when comparing the densities of various substances. All other substances will have **unique** densities that are either greater than or less than 1.0 g/cm^3.

EQUIPMENT

Check off the items to make sure you have the following equipment and materials. If an item is missing, obtain it from your teacher.

___ triple-beam balance

___ 10-mL graduated cylinder

___ four dropper bottles with:

mineral oil

glycerine

alcohol (coloring added)

water (coloring added)

PROCEDURES

You will be working in one of three teams. Each team will be responsible for determining the density of one liquid (alcohol, mineral oil, or glycerine). You will then share your results with the other two teams in your group.

1. Determine the mass of the graduated cylinder; then add exactly 1.0 mL of the liquid chosen by your group. **NOTE:** Some of these liquids are very "sticky" and take a long time to run down the sides of the cylinder. To avoid this situation, you should slowly add drops of the liquid down the center of the cylinder.

	Volume of Liquid		
	1.0 mL mineral oil	1.0 mL gylcerine	1.0 mL alcohol
Mass of graduated cylinder with liquid	__.__ g	__.__ g	__.__ g
– Mass of cylinder itself	– __.__ g	– __.__ g	– __.__ g
Mass of liquid	__.__ g	__.__ g	__.__ g

The density of mineral oil is $1.0 \text{ mL } \overline{) \text{ g}} \quad \overset{.\quad \text{g/cm}^3}{}$

The density of glycerine is $1.0 \text{ mL } \overline{) \text{ g}} \quad \overset{.\quad \text{g/cm}^3}{}$

The density of alcohol is $1.0 \text{ mL } \overline{) \text{ g}} \quad \overset{.\quad \text{g/cm}^3}{}$

2. Compare your results with those of the other two teams in your group. Which team's cylinder holds the liquid with the greatest density? To that cylinder, add the other two liquids and the water from the dropper bottle. **NOTE:** The liquids must be added one at a time and in an order of decreasing densities. The liquid with the least density is added last. **NOTE:** Add the liquids slowly down the tilted side of the graduated cylinder that contains the liquid with the greatest density.

NOTE: Do **NOT** shake the liquids in the cylinder.

3. Complete the figure based upon your observations.

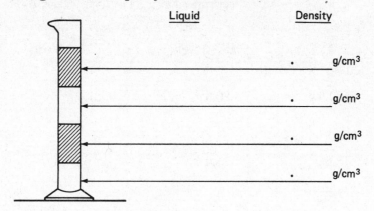

Liquid	Density
	. _____ g/cm³
	. _____ g/cm³
	. _____ g/cm³
	. _____ g/cm³

TEST YOUR UNDERSTANDING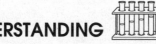

1. Even if you didn't have the information from your measurements, how could you tell which liquid has the least density after all four liquids are added to one graduated cylinder?

    _____

    _____

    _____

2. The fresh water of a river that empties into an ocean is often found floating above salt water far out to sea. What hypothesis can you form that might explain this **phenomenon**?

    _____

    _____

3. When oil from a fuel tanker spills into the ocean, the oil can sometimes be recovered by engineers using equipment that scoops up the oil as it floats on the surface. What hypothesis can you form to explain this phenomenon?

    _____

    _____

    _____

GOING FURTHER

Four test tubes contain clear, colorless, and odorless liquids. Use the following information to identify the test tube(s) that probably contain(s) water:

	Volume of Liquid in Test Tubes A–D			
	(A) 1.0 mL	(B) 2.0 mL	(C) 3.0 mL	(D) 4.0 mL
Mass of test tube with liquid	20.0 g	20.6 g	21.6 g	23.0 g
Mass of test tube itself	18.6 g	18.6 g	18.6 g	18.6 g

4. Which test tube(s) contain(s) water?

    _____

5. How did you arrive at this conclusion?

    _____

    _____

    _____

Refer back to question 3 (in "Test Your Understanding"). How would you design an experiment to test the hypothesis you formed about oil and salt water?

6. What equipment would you need?

    _____

    _____

    _____

7. What steps would you take and why?

    _____

    _____

    _____

    _____

8. How would you know if your predicted results were correct?

    _____

    _____

22

Estimating Densities

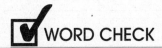 WORD CHECK

estimate	a rough, or approximate, calculation of something's value

PROBLEM: How do you estimate the density of a substance that is difficult to measure?

Estimation is a valuable technique in science, for scientists often take known facts and use them to gain additional information by making **estimates**.

For example, on another worksheet you made measurements of the volume and mass of four different liquids and determined that their densities were as follows:

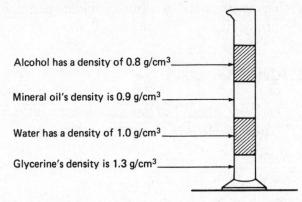

Alcohol has a density of 0.8 g/cm³

Mineral oil's density is 0.9 g/cm³

Water has a density of 1.0 g/cm³

Glycerine's density is 1.3 g/cm³

In today's work, you will use these facts to estimate the density of substances that are more difficult to measure.

EQUIPMENT

Check off the items to make sure you have the following equipment and materials. If any item is missing, obtain it from your teacher.

___ four dropper bottles with: water, mineral oil, alcohol, and glycerine

___ 10-mL graduated cylinder

___ egg-white chunks   ___ ice

PROCEDURES

Use the medicine droppers to add the four liquids, one at a time, to a 10-mL graduated cylinder. Note the following precautions:

- Use no more than about 2.0 mL of each liquid.
- Make sure you add the liquids in the correct order of decreasing densities. The liquid with the least density is added last.

1. Add a small chunk of hard-boiled egg white to the cylinder. What layer(s) does the egg white pass through?

   _____

   _____

2. What layer does the egg white stop on top of?

   _____

3. Use your observations to draw the position of the egg white on the picture of the graduated cylinder at the beginning of this worksheet. Label it.

4. Add a small chunk of ice to the cylinder. What layer(s) does it pass through?

   _____

5. What layer does the ice stop on top of?

   _____

6. Use your observations to draw the position of the ice chunk on the picture of the graduated cylinder. Label it.

TEST YOUR UNDERSTANDING

1. What can you now estimate about the density of egg white? The density of egg white is ____ (Write the letter of the correct choice.)
   a. less than 0.8 g/cm^3.
   b. between 0.8 and 0.9 g/cm^3.
   c. between 0.9 and 1.0 g/cm^3.
   d. between 1.0 and 1.3 g/cm^3.
   e. more than 1.3 g/cm^3.

2. On what information did you base your answer to question 1?

   _____

   _____

   _____

   _____

   _____

3. What can you now estimate about the density of ice? The density of ice is ___ (Write the letter of the correct choice.)
    a. less than 0.8 g/cm^3.
    b. between 0.8 and 0.9 g/cm^3.
    c. between 0.9 and 1.0 g/cm^3.
    d. between 1.0 and 1.3 g/cm^3.
    e. more than 1.3 g/cm^3.

4. On what information did you base your answer to question 3?

    _____

    _____

    _____

    _____

5. Why do ice cubes float in a glass of water?

    _____

    _____

GOING FURTHER

When a fresh egg is placed in a beaker of water, the egg settles to the bottom. If more and more salt is then added to this beaker, the egg slowly rises until it is floating on top of the water.

6. Why does the egg sink to the bottom of the water at the start?

    _____

    _____

7. How must the water have been changed by the addition of the salt?

    _____

    _____

    _____

8. On an earlier worksheet you determined the volume of blocks and cubes by measuring their height, length, and width with a ruler. Why can't you use the same technique with a chunk of egg white or ice?

_____

_____

CHAPTER
23

Determining the Density of a Regularly Shaped Solid

☑ WORD CHECK

regularly shaped	an object that is formed or built according to some established rule, standard, or pattern

PROBLEM: How do you determine the density of a cube or block?

All solids have both a volume and a mass as do liquids. Therefore, solids also have definite densities.

In your work today, you will combine measurement techniques you have used before to determine the densities of different solids that are shaped as blocks or cubes.

EQUIPMENT

Check off the items to make sure you have the following equipment and materials. If any item is missing, obtain it from your teacher.

___ ✓ metric ruler   ___ ✓ triple-beam balance

___ ✓ assorted blocks and cubes of the same shape and size

PROCEDURES

- Recall that the volume of a block or cube is determined by multiplying height × width × length.

- Recall that all three of these measurements are equal for a cube.

- Recall that density is a comparison of mass divided by volume, or $D = \dfrac{mass}{volume}$

1. Determine the mass and volume of each of two blocks (A and B) made from different materials but that have the same shape and size.

2. Use the measurements to complete the table. Use the nearest whole numbers to make things simpler (for example, use 2.0 instead of 2.2, or 2.0 instead of 1.8).

	Block A	Block B
Material	~~steel~~ metal ~~steel~~	copper
Height	2~~4~~.7 cm	2.5 cm
Length	2~~4~~.7 cm	2.5 cm
Width	24.9 cm	2.5 cm
Volume	~~20~~. ___ cm³	25 ~~14000000~~ cm³
Mass	6. ___ g	50 ~~XX~~ g
Density	1. ___ g/cm³	6. ___ g/cm³

3. Below are blocks of different volumes and masses. Use your metric ruler to measure and determine their volumes. The mass of each block is given. Calculate the density of each block from this information.

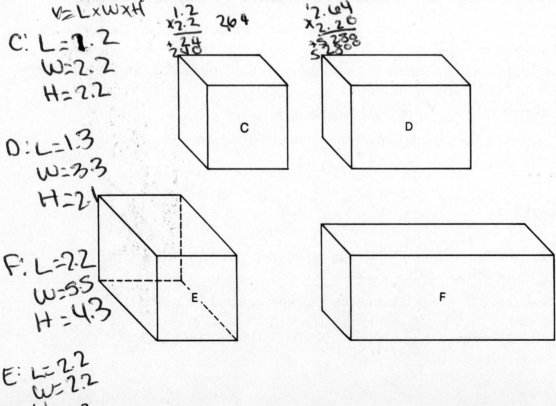

V = L × W × H

C: L = 2.2
   W = 2.2
   H = 2.2

1.2
×2.2   264
────
~~24~~

2.64
×2.20
──────
~~288~~
5~~2800~~

D: L = 13
   W = 3.3
   H = 2

F: L = 2.2
   W = 5.5
   H = 4.3

E: L = 2.2
   W = 2.2
   H = 2.2

$D = \frac{M}{V}$

	Block C	Block D	Block E	Block F
Volume	5.808 cm³	9.009 cm³	10.648 cm³	52.0 cm³
Mass	77.2 g	67.8 g	154.4 g	113.0 g
Density	_____ g/cm³ 13.30	_____ g/cm³ 7.53	_____ g/cm³ 14.5	.450 g/cm³ 2.17

TEST YOUR UNDERSTANDING

1. Blocks A and B are the same size and shape, and yet one feels heavier than the other. How is this possible?

    the one that feels heavier is denser and has a greater mass

    The following questions refer to blocks C, D, E, and F.

2. Which of these blocks is a cube?

    C E

3. How can you explain your answer to question 2?

    the length, width, and height are equal

    The density of lead is 11.3 g/cm³.

    The density of gold is 19.3 g/cm³.

4. Which blocks are made of lead?

    E and C

5. Which blocks are made of gold?

    D

$d = \frac{M}{V}$ $11.3 = \frac{x}{1}$

GOING FURTHER

6. What would be the mass of a cube of lead that measures 1.0 cm in height?   **11.3**

7. What would be the mass of a cube of gold that measures 1.0 cm in length?   **19.3**

C H A P T E R

24

Determining the Density of an Irregularly Shaped Solid

☑ WORD CHECK

irregularly shaped	having uneven surfaces rather than even sides

PROBLEM: How can you determine the density of a solid that has uneven sides?

Many years ago, a rich and powerful king was given a beautiful crown as a gift by people who wanted to win the king's friendship. The king was pleased, but he was not sure that the crown was really made of solid gold as the gift-givers claimed. He suspected that the crown might have been made of a cheaper metal, such as lead, and then covered with just a thin layer of gold.

To solve this puzzle, the king sent for the scientist Archimedes. He told Archimedes what he wanted to find out and cautioned him to perform an analysis in such a way that the crown would not be taken apart or damaged.

In your work today, you will be following procedures similar to those Archimedes used in finding his answer.

EQUIPMENT

Check off the items to make sure you have the following equipment and materials. If any item is missing, obtain it from your teacher.

- 10-mL graduated cylinder
- solids of different shapes and sizes
- dropper bottle with water
- triple-beam balance

Determining the Density of an Irregularly Shaped Solid

PROCEDURES

On another worksheet, you determined the volume of a solid that had uneven sides. By using a graduated cylinder, you measured the amount of water that was pushed out of the way, or displaced, by an **irregularly shaped** solid; this was the solid's volume.

1. First, find the mass of each solid.

2. Then, add each solid separately to a fixed volume of water in a graduated cylinder. Recall that the difference between the first and second readings of the volume of water in the cylinder (before and after the solid has been added) is the volume of the solid.

3. Enter your data for each solid in the table. Use your data to determine each solid's density.

	Solid A _ball steel_	Solid B _cylinder_	Solid C
Mass	63.17 g	5.01 g	150.0 g
Volume	6.44 mL	.6 mL	25.0 mL
Density	1.0 g/cm³	8.35 g/cm³	6.0 g/cm³

TEST YOUR UNDERSTANDING

1. Why do you first determine the mass of the solid before adding it to the water in the graduated cylinder?

   the solid might be wet which could increase its mass

2. The solids you have just measured all have different masses and volumes. What hypothesis might you form about these three different solids if you determined that they all had the *same* density?

   they would all either sink, float, or be neutally bayent

3. What hypothesis might you form if you determined that they all had *different* densities?

   some might sink while the others float or have neutral boyancy

4. Now let's get back to Archimedes and his problem with the crown. Assume that he carried out the same types of procedures you did and discovered the following:

   Mass of the crown 650 g

   13 g/cm³

   Volume of the crown 50 mL

Would the king have the gift-givers arrested or rewarded? Why? (Recall that the density of lead is 11.3 g/cm^3, and the density of gold is 19.3 g/cm^3.)

arrested - the density is 13 g/cm^3 so it was not made of solid gold.

5. Suppose Archimedes had made the following measurements:

    Mass of the crown 965 g

    Volume of the crown 50 mL

    Would the king now have the gift-givers arrested or rewarded? Why?

    rewarded the crown's density is greater than 19.3 g/cm^3

GOING FURTHER

6. You can usually tell if a coin is pure silver or if it is made of a combination of copper and silver. Simply look at the edge of the coin to see if a copper color is visible. (If you could not tell from the edge of the coin, what two things could you measure to find out if the coin is a combination of metals or pure silver? Why?)

    density and mass volume mass because then you would be able to calculate the coin's density

C H A P T E R

25

Testing a Hypothesis

 WORD CHECK

temperature	how hot or cold something is as measured with a thermometer
thermometer	an instrument that has a glass bulb attached to a fine tube of glass, with a numbered scale, that contains a liquid, which is sealed in and rises and falls with changes of temperature

PROBLEM: How accurately is your body able to determine differences in temperature?

• Before giving the bottle to her infant, a mother spills a few drops of milk from her baby's bottle onto her forearm.

• Before stepping into the bathtub, a young child carefully puts a toe into the water.

- When she thinks her child has a fever, a mother places her lips to her child's forehead.
- When checking to see if an iron is hot enough to use on his shirts, a young man wets his fingers before touching the bottom of the iron.

In each of these everyday situations, people are using different parts of their body in an attempt to check **temperatures**. Whether or not this can be done accurately is what you will be testing today. What is your hypothesis?

EQUIPMENT

Check off the items to make sure you have the following equipment and materials. If any item is missing, obtain it from your teacher.

___ two small cups   ___ one large container

PROCEDURES

1. Do the following—both at the *same time*.

   Place the index (or pointer) finger of your *left* hand into a cup of very *warm* water for 30 seconds. Place the index (or pointer) finger of your *right* hand into a cup of very *cool* water for 30 seconds.

Warm water
left hand

Cool water
right hand

2. Have your partner record your observations for you on your worksheet.
   a. The finger that was in the warm water (left hand) feels (warm, cool)

   _____

   b. The finger that was in the cool water (right hand) feels (warm, cool)

   _____

3. Now pour the water from both cups into one larger container. Place both index fingers into this water at the same time for 30 seconds.

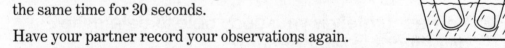

4. Have your partner record your observations again.

   a. The finger that was in the warm water (left hand) now feels (warm, cool)

   _____

   b. The finger that was in the cool water (right hand) now feels (warm, cool)

   _____

5. Next, have your partner repeat the experiment while you record the observations on your partner's worksheet.

TEST YOUR UNDERSTANDING

1. Why are you puzzled by the observations you made when both fingers were in the same container of water at the same time?

   _____

   _____

   _____

2. How would you answer the original problem posed at the beginning of this worksheet? Base your answer upon the results of your experiment.

   _____

   _____

   _____

3. What does your ability to detect changes in temperature sometimes depend on?

   _____

   _____

   _____

   _____

4. Why are **thermometers** important tools in science?

   _____

   _____

   _____

GOING FURTHER

5. When a person who has been outside in the bitter cold enters a room, he or she might complain that the room is too warm. At the same time, a person who has been in that same room for several hours might consider the room to be too cool. Why is this possible?

   _____

   _____

   _____

6. Why was it important, in today's experiment, to make sure that both fingers were placed in the water at the same time and for equal amounts of time?

_____

_____

_____

C H A P T E R

26

Making Observations

WORD CHECK

expansion	an increase in the volume of an object
contraction	a decrease in the volume of an object
rigidly	appearing stiff and unyielding, or firmly inflexible

PROBLEM: How do the volumes of solids, liquids, and gases change as they are heated or cooled?

Trying to determine temperature changes by using parts of your body is not very reliable. This explains why the thermometer is needed in scientific work.

By using today's worksheet and by observing the demonstrations performed by your teacher, you will gain an understanding of how thermometers work. However, examine the following questions and descriptions before you begin. Their explanations will help you in answering today's problem.

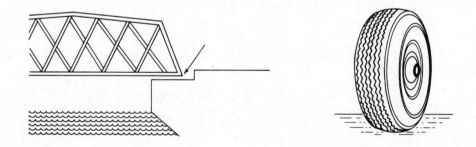

- Why are sidewalks and concrete highways built with spaces between each section instead of being formed as one continuous piece?

- Why are bridges built resting in place on their foundations rather than being **rigidly** attached?

- Why do long-distance truckers let some air out of their tires before starting out in hot weather?

107

PROCEDURES

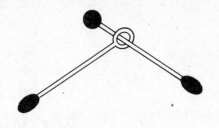

The equipment being used in this demonstration consists of a brass ring and a brass ball—both at the end of long handles. Note how easily the ball passes through the ring.

1. Heat the ball in a flame, and then try to pass the ball through the ring. Describe what happens.

   _____

   _____

2. Cool the ball in water. Then try to pass it through the ring. What do you observe?

   _____

   _____

3. With the ball inserted through the ring, heat both the ball and the ring in the flame. Once again try to pass the ball through the ring. What happens this time?

   _____

   _____

4. Cool both the ring and the ball in water. Now try to separate them. What happens?

   _____

   _____

The term **expand** means to become larger in volume. The term **contract** means to become smaller in volume. Answer the following questions based upon your observations in 1–4. Use the terms expand (expansion) and contract (contraction).

5. How do solids usually change when they are heated?

   _____

6. How do solids usually change when they are cooled?

   _____

Answer the next group of questions based upon the observations you make when air is heated and cooled.

7. The air in the flask in the setup at the right is heated until a change takes place in the balloon. What is this change?

   _____

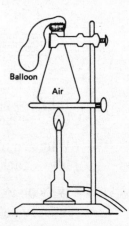

Balloon

Air

8. The air in the flask is then allowed to cool. What change occurs in the balloon?

   _____

9. How must the air inside the flask have changed as the air became hotter? How did you arrive at this conclusion?

_____

_____

_____

10. How must the air inside the flask have changed as the air was cooled? How did you arrive at this conclusion?

_____

_____

_____

Answer the next group of questions based upon the observations you make when a liquid is heated and cooled.

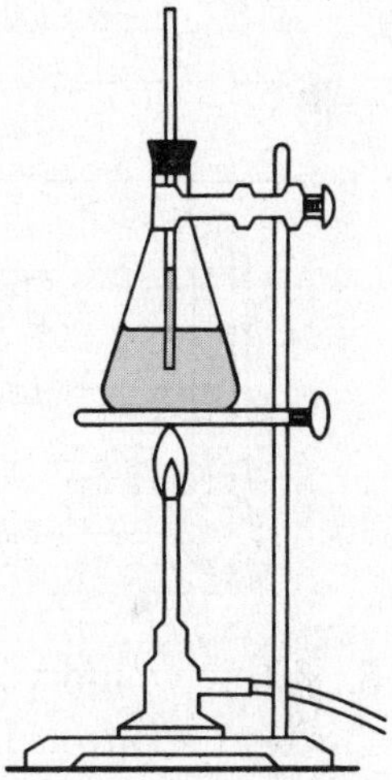

11. As water in the flask illustrated at the right is heated, the level of the water in the tube changes. How does the level change?

_____

12. The water is now allowed to cool. What change do you observe?

_____

_____

13. How must the water in the flask have changed when the water was heated? cooled?

_____

_____

_____

TEST YOUR UNDERSTANDING

1. How do solids, liquids, and gases generally change when they are heated or cooled?

_____

_____

_____

2. Why must sidewalks and concrete highways be built with spaces between their sections?

_____

_____

_____

3. Why is it important for a bridge to be built with spaces between the bridge and its foundation?

_____

_____

_____

4. Why do truck drivers let some air out of their tires before traveling in hot weather?

_____

_____

_____

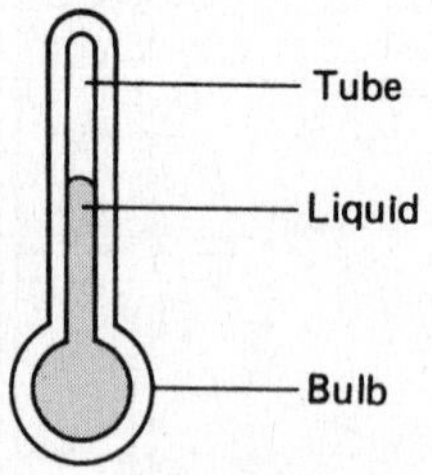

Inside a typical thermometer, as illustrated at the left, is a narrow tube that is partially filled with a liquid. You can see that the tube is connected at the base to a bulb that serves as a container of additional fluid.

5. How would you expect the height of the liquid in the tube to change as the temperature increases? Why?

_____

_____

6. How would you expect the height of the liquid to change as the temperature decreases? Why?

_____

_____

_____

7. The thermometer in the figure is missing a scale. Why is a scale an important part of a thermometer?

_____

_____

_____

GOING FURTHER

8. When a hand is placed over the glass bulb, as shown in the figure below, the air inside the bulb becomes warmer. What do you predict will happen to the level of the water in the tube? Why?

_____

_____

_____

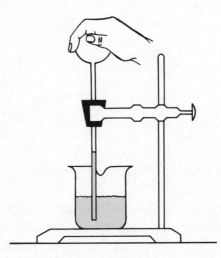

9. The equipment used in question 8 is called an *air thermometer*. How are an air thermometer and a fluid-filled thermometer similar in the way in which they work?

_____

_____

_____

_____

27

Summarizing Observations; Drawing Conclusions

✔ WORD CHECK

freezing point	the temperature at which a liquid turns into a solid
boiling point	the temperature at which a liquid turns into a gas

PROBLEM: How do you use a thermometer?

Recall what you learned earlier. In general, accurate measurements of temperature are not possible when parts of the body are used in determining temperature. Also, recall your observations of how a thermometer works as the liquid inside expands or contracts during temperature changes.

The thermometer you will be using today contains alcohol to which a red color has been added. An alcohol thermometer is not quite as accurate as a mercury thermometer, but it is easier to read and is suitable for the work you will do.

The units on the scale of your thermometer are known as *degrees Celsius*. The key points marked on this scale are the **freezing point** of water (0°C) and the **boiling point** of water (100°C).

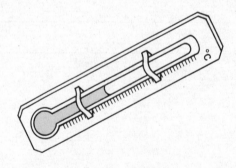

EQUIPMENT

Check off the items to make sure you have the following equipment and materials. If any item is missing, obtain it from your teacher.

___ alcohol thermometer

___ dropper bottle of alcohol

___ dropper bottle of water

___ cotton

PROCEDURES

- **CAUTION:** Do *not* shake the thermometer to lower the level of the liquid, for you may break it.
- Do *not* hold the thermometer by the bulb because you may cause a high reading.

1. Record the temperature of the classroom.

   _____ °C

2. Gently place your thumb and first finger of one hand against the bulb of the thermometer. Record the final temperature reading.

   _____ °C

3. Hold a water-dampened piece of cotton against the bulb. From a distance of about 12 cm, blow on the cotton, and then record the temperature at the end of one minute.

   _____ °C

4. Warm the thermometer with your fingers. Then dampen a piece of cotton with alcohol, and hold it against the bulb. Blow against the cotton again as you did with the water. Record the results.

   _____ °C

5. Place all your results into one table. This will make it easier to summarize your observations and to arrive at conclusions.

	Temperature
Air	_____ °C
Fingers	_____ °C
After blowing on cotton soaked with water	_____ °C
After blowing on cotton soaked with alcohol	_____ °C

TEST YOUR UNDERSTANDING

1. How does the temperature of your fingers compare with the temperature in the classroom? (higher, lower, about the same)

2. When you blow on a surface that has been moistened with a liquid, such as water or alcohol, the liquid seems to dry up and disappear. *Evaporation* has taken place as the liquid changes to a gas. How did the temperature readings change after you blew on the cotton soaked in water?

_____

GOING FURTHER

3. Suppose you had just come out of a swimming pool. Why would you feel chilled on a hot, windy day?

_____

_____

_____

4. When a patient has a high fever, the nurse may use an alcohol sponge bath to lower the patient's temperature. Why does this work?

_____

_____

C H A P T E R

28

Organizing Data

✓ WORD CHECK

tallying	counting items by checking them off

PROBLEM: How can information be organized into useful form?

Students in a science class were curious to know how well they had done on a recent test in comparison with another science class taught by the same teacher. At the students' request, the teacher provided factual information called *data*. However, students would have to draw their own conclusions from the data. See what conclusions you arrive at after comparing the data from the two classes.

PROCEDURES

- The information provided by the teacher, which had not been organized in any way, is called *raw data*. The raw data (test scores) for the two classes were as follows:

The "Curious" Class								The "Other" Class							
72	75	90	70	77	65	54	78	98	81	75	66	47	89	82	69
72	69	73	89	64	73	70	87	70	63	66	87	76	69	72	82
66	57	84	93	74	65	96	77	84	70	75	77	68	65	56	71
79	82	71	80	79	62	67	59	75	79	79	85	76	72	79	74

- The students decided to organize the raw data into groups that are similar to the way marks are recorded as letter grades on their report cards. Example:

90-100 = A	65-69 = D
80-89 = B	Below 65 = F (failure)
70-79 = C	

- Scan the data for test scores that match each letter grouping. In the table below, record each count by using this method of **tallying**: ⫴⫼ Then place the total for each grade as a number at the end of each line. For example, one grouping has already been done for the "curious" class.

		The " Curious" Class		The "Other" Class
90s	As	3 ııı		1
80s	Bs	5 ⫴⫼		7
70s	Cs	14 ⫴⫼⫴⫼ ıııı		15
65+	Ds	5 ⫴⫼		6
64–	Fs	⫴⫼ ▸ 5		3

q – ııı
8 – ⫴⫼
7 – ⫴⫼⫴⫼ııııı
65 – ⫴⫼ı
64 – ⫴⫼ı

9 · ı
8 – ⫴⫼ıı
7 – ⫴⫼⫴⫼⫴⫼
65 – ⫴⫼ı
69 – ıı ı

1. There were **32** students in each class. How can you check to see if you missed any test scores when you were placing them into letter groupings?

   add your tallys togethe

   Use the letter for one of these three statements to answer questions 2 to 5.

   a. The number was greater for the "curious" class than for the "other" class.
   b. The number was greater for the "other" class than for the "curious" class.
   c. The number was the same for both classes.

2. How did the number of students who passed the test compare in both classes?

   a

3. How did the number of students who got marks of 90 or more compare in both classes?

   a

4. If marks between 65 and 79 are considered average test scores, how did the number of students who were doing average work compare in both classes?

   b

5. If marks between 80 and 100 are considered good/excellent test scores, how did the number of students who were doing this type of work compare in both classes?

   c

TEST YOUR UNDERSTANDING

1. Which of the following statements best describes what the students in the "curious" class learned from organizing the raw data (or test scores) for the two classes? (Write the letter of the correct choice.)   *a*

   a. The "curious" class had more students with top grades in the 90's, but it also had more failing students.
   b. The "curious" class had more students who had top grades in the 90's, and it also had fewer students who failed.
   c. The "curious" class had fewer students with top grades, and it also had more students who failed.
   d. The "curious" class had fewer students with top grades, but it also had fewer students who failed.

2. When comparing their test scores, why were the students in the "curious" class wise in selecting another class that was the *same* type as theirs and that was taught by the *same* teacher?

   to make sure their experiment would be more accurate

GOING FURTHER

There is more than one way in which the raw data for the test scores could have been organized. For example, marks of A+, B+, and C+ would provide more exact information than just A, B, and C. Assume the following grade scale:

A+ = 95–100	A = 90–94
B+ = 85–89	B = 80–84
C+ = 75–79	C = 70–74

- Go back to the raw data (test scores), and use the information to complete this second grouping.

Handwritten notes in left margin:
95-100 ┌ 1
85-89 ─ 11
75-79 ─ ₭Ⅱ 1
90-94 - 11
80-84 - 11
70-74 ─ ₭Ⅱ 1
₭Ⅱ
85-89
75-79
100-95-1
95-90-
85-85- 111
84-80-1111
79-75- ₭Ⅱ 1111
74-70- ₭Ⅱ 1

		The "Curious" Class		The "Other" Class	
A+	95–100	B	1	l	1
A	90–94	ll	2		0
B+	85–89	ll	2	llB	3
B	80–84	lll	3	llll	4
C+	75–79	₭Ⅱ l	6	₭Ⅱ llll	9
C	70–74	₭Ⅱ lll	8	₭Ⅱ l	6

Use the letter for one of these three statements to answer questions 3 to 5.

   a. The number was greater for the "curious" class than for the "other" class.

   b. The number was greater for the "other" class than for the "curious" class.

   c. The number was the same for both classes.

3. How did the number of students who received scores that are classified as A's compare in both classes?

                a

4. How did the number of students who received scores classified as A+ compare in both classes?

                c

5. How did the number of students who received scores classified as C+ compare in both classes?

                b

CHAPTER
29

Recording Data

PROBLEM: How does the temperature of water change as ice melts in it?

You have learned that factual information, such as measurements scientists collect when they make observations, is called *data*. Data must be recorded as an experiment is going on, and the data should then be organized. When presented in neat, logical form, the data can more easily be examined and understood. A scientist should be able to summarize data, to use data in describing what took place during an experiment, and to reach a conclusion.

Today, you will collect and record data as you perform an investigation on how water temperature changes as ice melts in it. Then, you will examine the data to see what you can learn from your collection of facts.

EQUIPMENT

Check off the items to make sure you have the following equipment and materials. If any item is missing, obtain it from your teacher.

_____ student thermometer

_____ paper cup

_____ saucer or bottom of a petri dish

_____ tap water

_____ graduated beaker, 250 mL

_____ watch or clock with second hand or a stopwatch

_____ ice cube

PROCEDURES

1. Set up the equipment as shown in the figure. Add 100 mL of water to the cup.

2. When the thermometer has adjusted to the temperature of the water, record the reading in the data table that follows.

3. Put one ice cube in the water. One minute later, read and record the water temperature. You will continue to record the water temperature at one-minute intervals for the next 25 minutes or so. To work efficiently, one partner can make the readings on the thermometer while the other partner keeps track of the time and enters the readings in the table. Using a stopwatch may be helpful.

4. Mark the table by circling the time when the ice cube has melted completely, but continue to check and record the temperature until the table has been completed.

Data Table					
Water temperature (without ice)	6.2 °C				
→ Ice added ←		8 minutes	6.1 °C	18 minutes	6.1 °C
Water temperature (with ice)		9 minutes	6.1 °C	19 minutes	6.1 °C
		10 minutes	6.1 °C	20 minutes	6.1 °C
after 1 minute	6.7 °C	11 minutes	6.1 °C	21 minutes	6.2 °C
2 minutes	6.3 °C	12 minutes	6.1 °C	22 minutes	6.2 °C
3 minutes	6.1 °C	13 minutes	6.1 °C	23 minutes	6.3 °C
4 minutes	6.1 °C	14 minutes	6.1 °C	24 minutes	6.3 °C
5 minutes	6.0 °C	15 minutes	6.1 °C	25 minutes	7.3 °C
6 minutes	6.0 °C	16 minutes	6.1 °C	26 minutes	7.7 °C
7 minutes	6.0 °C	17 minutes	6.1 °C	27 minutes	7.7 °C

Examine your results; then, answer the following questions.

5. How did the temperature of the water change after you added the ice cube?

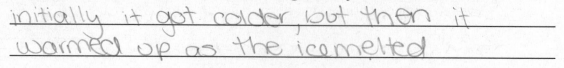

initially it got colder, but then it warmed up as the ice melted

6. When had the ice cube melted completely?

it took longer than 27 mins for it to melt

7. How did the temperature of the water change after the ice cube had melted completely?

it didn't melt completely in the 27 minutes

8. How did the final reading of water temperature compare with the reading before you added the ice cube?

it was higher

TEST YOUR UNDERSTANDING

1. Why did all students begin with the same volume of water?

to make sure their tests were the same

2. Why was it necessary to record the water temperature throughout the experiment instead of taking temperature readings only at the beginning and the end?

because the temperature of the water decreased and then it increased.

3. How might your results have differed if you had used a **Styrofoam** cup or an **insulated** cup instead of a paper cup? Explain your answer.

the water temperature would have been lower + the ice would have taken longer to melt b/c the cup would hold the temp. longer

4. If you and another team had been equally careful in making observations and collecting data, what is one factor that might have caused the two teams to obtain different readings?

the size of the ice cube or the air temperature in the room

GOING FURTHER

5. How might your data have been different if you had added two ice cubes instead of one to the water at the start of the experiment? What is your hypothesis?

    I think that having two ice cubes
    in the cup of water would have taken
    longer to melt b/c the water would
    have been colder

6. How could you find out if your hypothesis is correct? Write a plan for testing your hypothesis and show it to your teacher. If your plan is approved, you may want to carry out the experiment again.

    1. get 200mL of water in a graduated cylinder
    2. add 2 ice cubes
    3. record water temp every minute

Constructing a Line Graph

☑ WORD CHECK

line graph	a *graph* on which points representing numerical data are connected to form a line (The rise or fall of the line shows the increase or decrease between numerical data. A *line graph* is often used to show how something that is being measured changes over a period of time. It will show at a glance what is more likely to happen next.)

PROBLEM: What are some ways of using a line graph to understand data?

After collecting data, scientists organize the facts in different ways to see if they can learn more than may be evident from a data table. Very often, they construct a **line graph** from the data. Such a graph shows clearly the relationship between two factors, such as the changes from year to year in the population of a country. Line graphs also are a means for making comparisons and developing predictions.

Today, you will construct a line graph, using the data you collected as an ice cube melted in water.

PROCEDURES

A. Study the grid on which you will plot the data you have collected. First, read the title above the grid. It tells you that when the graph is completed, it will show the temperature changes that took place in water as an ice cube melted in it over a period of time.

B. Read the label along the left side of the grid. It tells you that the numbers reading from the bottom to the top are temperatures in degrees Celsius. Studying the numbers, you can see that the scale increases by units of 1 degree from 0°C at the bottom to 25°C at the top of the grid.

C. Read the label at the bottom of the grid. It tells you that the numbers represent time in 1-minute units. Reading from left to right, time increases from 1 minute to 27 minutes.

D. You and your partner will now plot your data on the grid. One partner can call out the time and temperature readings from the data table while the other partner marks the grid.

E. Begin by showing the water temperature before you added the ice. To do so, move a pencil lightly upward on the line labeled **Beginning water temperature (before ice was added)** until you reach the line with the temperature you recorded in your data table. Make a dot where the two lines cross (time and temperature).

F. Next, move to the line labeled *1 minute* and move up this line until you reach the cross line labeled with the temperature reading you made one minute after you added the ice cube to the water. Make a dot where the 1-minute line and the temperature line cross each other.

G. Continue in the same way, making dots to show all the readings you collected over the 27-minute period of recording temperature.

H. Use a line to connect the dots.

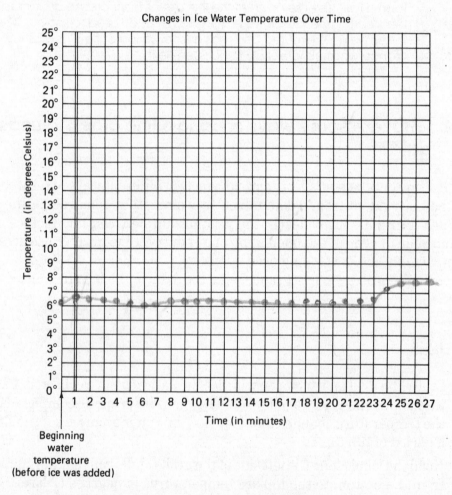

I. Study the graph to see what you can learn from it.

1. Study the line graph and find the point representing the lowest temperature the water reached. What was the lowest temperature?

6 °C

2. What happened to the temperature of the water in the few minutes after the lowest temperature had been reached?

it began to increase

3. Check your data table to find the mark you made to show when the ice cube had melted completely. How many minutes after you added the ice to the water did the ice melt completely?

it took longer than 27 mins for the ice cube to melt

4. How does the time it took for the lowest temperature to be reached (see question 1) compare with the time it took for the ice cube to melt completely (see question 3)?

It took longer than

TEST YOUR UNDERSTANDING

1. Why is it an advantage to show the data you collect in an experiment in the form of a line graph?

line graphs are very useful when you are displaying data that has been recorded over a period of time.

2. If you had continued the experiment by recording temperature readings for another half hour, what hypothesis would you have formed about the direction the line of your graph would have taken?

the water's temperature would have continued to increase

3. Which observations and data support the hypothesis you described in answer to question 2?

after the ice cube melts the water's temperature began to increase gradually

GOING FURTHER

The units marked off on a grid for a graph do not always change by single numbers. They may increase by 2's or 100's, for example. When you plot data on such a grid, you sometimes have to estimate approximately where to place the marks showing the data you have collected. You will now construct a line graph on a grid requiring estimations.

Reported Cases of Disease A
1996-2000

Year	Number of reported cases*
1996	1300
1997	2800
1998	5500
1999	9500
2000	10,000

*Numbers rounded off to the nearest 100.

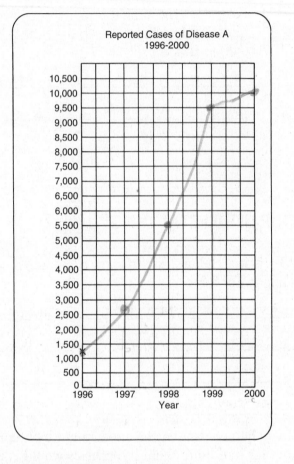

4. The data table shows the number of reported cases of Disease A in the United States from 1996–2000. Plot the information from the data table onto the grid, and connect the marks to make a line graph. The information for the first year has been marked with an X as an example.

5. Study the line graph you just constructed. What pattern of change in reported cases of Disease A between the years 1996 and 2000 does the line graph reveal?

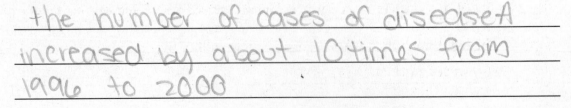

the number of cases of diseaseA increased by about 10 times from 1996 to 2000

31

Constructing and Interpreting a Double-Line Graph

WORD CHECK

decade	a period of 10 years
life expectancy	the number of years an average person born in a certain year is expected to live

PROBLEM: How can you construct a line graph to show two sets of data?

Sometimes, scientists have two sets of data they want to compare. By plotting the related sets of data on a single grid, they can obtain two line graphs that enable them to make a comparison more easily. A double-line graph is formed when both sets of data are plotted on the same grid.

Today you will use this technique to plot and compare two sets of related data. Then, by examining the double-line graph showing two sets of information at the same time, you will be able to make comparisons, draw conclusions, and make predictions.

PROCEDURES

First, look at the data table on the next page. It shows **life expectancy** in the United States at the beginning of each **decade** from 1900 to 2000. Life expectancy is the number of years babies born in a particular year can be expected to live. Checking the data table, you can see that male babies born in 1940 are expected to live 61 years.

Life expectancy is influenced by the general state of health and well-being of a country's population. The graphed data is, therefore, of interest to scientists and others who work to improve the conditions that affect average length of life.

1. Now, write a title above the grid that will accurately describe the data you will be showing in graph form.

2. Then, examine the grid to understand the units reading up the left side and across the bottom of the grid. What information does the left side of your grid describe?

life expectancy in years

What information does the bottom of your grid describe?

year (1900-2000)

3. Using the method you learned in Lesson 30, plot the life expectancy for males. Use a series of dashes as shown in the key to connect the dots.

4. Repeat the same process to plot the life expectancy for females. Connect these dots with a solid line as shown in the key.

Life Expectancy
in Years, 1900-2000*

Year	Male	Female
1900	46	48
1910	48	52
1920	54	55
1930	58	62
1940	61	65
1950	66	71
1960	67	73
1970	68	75
1980	70	78
1990	72	79
2000	74	80

*Numbers rounded off
to the nearest whole
numbers

KEY:

-------- Males
———— Females

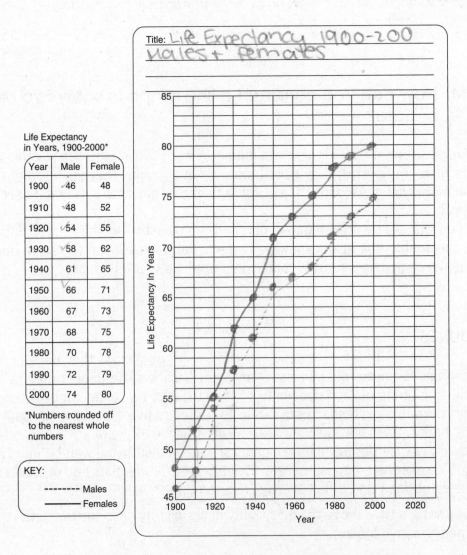

Title: Life Expectancy 1900-200 Males + Females

TEST YOUR UNDERSTANDING

1. Make a general statement that describes the changes in life expectancy for both males and females since 1900.

    both have increased

2. In 1900, how did the life expectancy for females compare with that for males?

    the females always had a greater life expectancy

3. In 2000, how did the life expectancy for females compare with that for males?

    females: 80 → females > males
    males: 74

4. If the pattern of change in life expectancy for females and males continues as it has in the past, you can predict that life expectancy will ___A___ (Write the letter of the correct choice.)

    a. continue to increase more for females than for males.
    b. continue to increase more for males than for females.
    c. continue to increase equally for both males and females.
    d. stay about the same for both males and females.

5. The year in which life expectancy was most alike for females and males was _____.

GOING FURTHER

Interpreting the Double-Line Graph

For each of the following items, choose the term or phrase that best completes the statement. Write the letter of the correct answer in the answer space.

6. In a large city in the United States, Disease B has been occurring at a rate that is different from its occurrence in the rest of the country. Using the information in the data table on page 130, construct a double-line graph. This will help you compare the rate at which this disease is appearing in the city versus the country as a whole.

    Plot the two sets of information according to the key. In addition, give the completed graph a name that describes the information the graph displays.

*Reported Cases of Disease B
1996-2000

Year	Number of Cases in New York	Number of Cases in the United States
1996	600	1,300
1997	1,000	2,800
1998	1,700	5,500
1999	2,500	9,500
2000	2,800	10,100

*Numbers have been rounded off

KEY
Reported cases of Disease B
——— City
– – – – United States

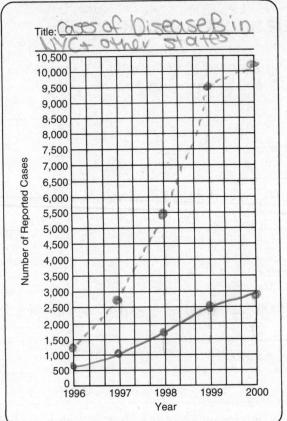

Title: Cases of Disease B in NYC + other states

7. Since 1996, Disease B has been _B_ (Write the letter of the correct answer.)
   a. decreasing in both the city and the United States.
   b. increasing in both the city and the United States.
   c. increasing in the city but decreasing in the United States.
   d. decreasing in the city but increasing in the United States.

8. The greatest increase in the occurrence of Disease B in the United States took place between the years _C_ (Write the letter of the correct answer.)

   a. 1996 and 1997
   b. 1997 and 1998
   c. 1998 and 1999
   d. 1999 and 2000

9. Disease B is ____ (Write the letter of the correct answer.)
    a. increasing at the same rate in the city as in the rest of the United States.
    b. decreasing at the same rate in the city as in the rest of the United States.
    c. increasing in the city faster than in the rest of the United States.
    d. increasing faster in the rest of the United States than in the city.

32

Interpreting a Bar Graph

☑ WORD CHECK

bar graph	a *graph* on which the data are shown as bars of different lengths that correspond to the numbers that they represent
circulatory diseases	*diseases* that affect the blood, blood vessels, or the heart

PROBLEM: What information can you learn from a bar graph?

The line graphs you constructed and interpreted in previous worksheets are good ways of showing changes over a period of time of something that is measurable, such as the spread of disease. You also learned how a double-line graph can be used to compare two sets of related data, such as the changes in life expectancy for males and females over the same period of time.

Today, you will learn another way to display and interpret data in the form of a graph—a **bar graph**. A bar graph is used to make comparisons among numbers of similar types of information or data that have no effect on one another. The numbers making up this data are used to determine the length of the bars on the graph.

PROCEDURES

1. On the graph that follows, what information do the labels provide at the base of the graph?

    <u>the different leading causes of</u>
    <u>diseases,</u>
    _____

2. Why is the information shown on the side of the graph important?

    <u>it shows the percent of all</u>
    <u>deaths from a certain disease</u>

Leading Causes of Death in
the United States In 1900

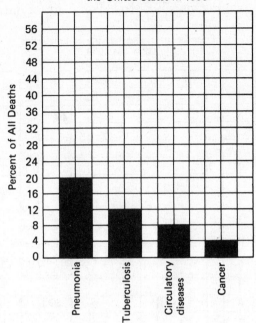

3. What was the major cause of death in 1900?

___pneumonia___

4. Which of the four diseases shown on the graph caused the smallest percent
of deaths in 1900?

___cancer___

5. What percent of all the deaths in 1900 were caused by **diseases** of the **circulatory**
system?

___8%___

6. What percentage of all the deaths in 1900 were caused by just these four diseases?

___44%___

7. The combined percent of deaths due to pneumonia and tuberculosis was _C_
(Write the letter of the correct choice.)

   a.   about the same as the combined percent of deaths due to circulatory diseases
and cancer.

   b.   less than the combined percent of deaths due to circulatory diseases and cancer.

   c.   almost three times greater than the combined percent of deaths due to circulatory diseases and cancer.

   d.   about five times greater than the combined percent of deaths due to circulatory
diseases and cancer.

TEST YOUR UNDERSTANDING

Use your understanding of bar graphs to interpret the following graph, which gives two
sets of data for each item. One set is the same as the data you have just studied—leading

causes of death in 1900. The other set shows the percent of all deaths due to these causes in 2000. The key beneath the graph shows how to tell one set of data from the other.

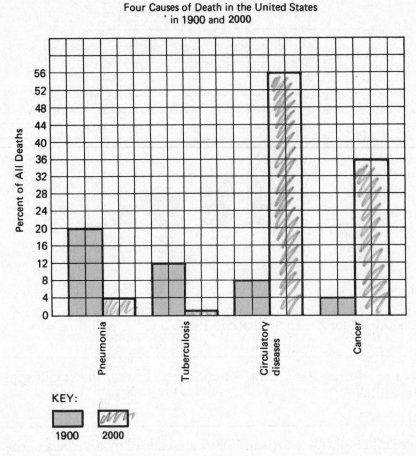

Four Causes of Death in the United States in 1900 and 2000

KEY: ☐ 1900   ▨ 2000

1. Of the four diseases shown, which disease caused the highest percent of deaths in 2000?

2. Of the four diseases, which disease caused the smallest percent of deaths in 2000?

   tuberculosis

3. What percent of all deaths in 2000 were due to cancer?

   _52%_

4. Choose the phrase that best completes the statement of how the percent of all deaths due to cancer compared in 1900 and 2000. Write the letter of the correct choice in the answer space.

   The percent of all deaths due to cancer was _____
   a. about the same in 1900 and in 2000.
   b. higher in 1900 than in 2000.
   c. about four times higher in 2000 than in 1900.
   d. about nine times higher in 2000 than in 1900.

5. To summarize the data shown in the double-bar graph, complete the following statement with one of the choices given below it.

Since 1900, the percent of all deaths due to pneumonia and tuberculosis has ___D___
(Write the letter of the correct choice.)

a. stayed about the same while the percent of all deaths due to circulatory diseases
    and cancer has increased.

b. decreased while the percent of all deaths due to circulatory diseases and cancer
    has remained about the same.

c. decreased while the percent of all deaths due to circulatory diseases and cancer
    has increased slightly.

d. decreased while the percent of all deaths due to circulatory diseases and cancer
    has increased greatly.

GOING FURTHER

6. The amount of energy provided by coal throughout the world each year is
    measured in units called quads. In 1975, coal supplied less than 100 quads of energy
    worldwide. About 600 quads were consumed in 2000. Energy from coal will
    probably rise to 1,200 quads in the year 2050 and to 2,300 quads by 2100. Would you
    use a linegraph or bar graph to diagram this data? Why?

    I would use a line graph because it could show the changes in coal usage more clearly than a bar graph would.

7. Studies of people who smoke cigarettes have shown that the risk of developing lung
    cancer is 59% for those who continue to smoke, 47% for those who are seriously
    attempting to stop smoking, and 35% for those who have successfully quit. Would
    you use a line graph or a bar graph to diagram this data? Why?

    I would use a bar graph to diagram this data because the information was not recorded over a period of time.

8. Why are both line graphs and bar graphs useful tools in science?

    Line graphs are good for showing data over a period of time, and bar graphs are good for comparing multiple groups.

C H A P T E R
33

Constructing a Bar Graph

☑ WORD CHECK

air pollution	the contamination of the air especially with waste materials produced by the human population
atmosphere	either the whole mass of air surrounding the earth or the air of a specific area (locality)

PROBLEM: How can you design a bar graph from data?

University scientists have issued a report based on research of average homes in six cities. The homes were studied to find out whether **air pollution** was worse indoors or outdoors.

In your work today, you will be assembling the report's data into tables. You will then use this information to construct bar graphs.

PROCEDURES

Your first set of data concerns air pollution *inside* the homes in these cities. Particles from dusting, cooking, and smoking pollute 20% of the air inside homes in city P. For city T, the figure is 22%; city K is 48%; city W is 28%; city St is 44%, and city S is 40%. Use this information to complete the data table.

Air Pollution Inside Homes	
City	Percentage of inside air polluted by particles in the home
P	20%
T	22%
W	28%
S	40%
St	44%
K	48%

Now use this information to plot a bar graph. The bottom of each bar has been started for you, and the bar for city P has been completed as a sample. Use shading for the bars showing the inside air pollution. ■

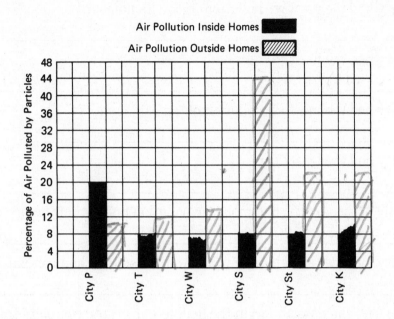

This next data table contains information that describes the percentage of air (*outside* homes) polluted by particles in the **atmosphere** in each of these cities. The cities are now arranged in the same order as in the previous data table.

Air Pollution Outside Homes	
City	Percentage of outside air polluted by particles in the atmosphere
P	10%
T	12%
W	14%
S	44%
St	22%
K	22%

Using the same graph, construct the bars for the pollution that is outside the homes in these cities. Use the columns to the left of the bars you have already completed. You will then have a double-bar graph that enables you to make comparisons between the indoor and outdoor air pollution in each city. Use cross-hatching for the bars showing the outside air pollution. ▨

TEST YOUR UNDERSTANDING

1. Which city had the most air pollution *inside* its homes?

   city P

   Which city had the least?

   city W

2. Which city had the most air pollution *outside* its homes?

   city S

   Which city had the least?

   city P

3. How many of these cities had more air pollution inside their homes than outside?

   1

4. Which city (or cities) had more air pollution outside its homes than inside?

   T, W, S, St, K

5. To summarize the data shown in the double-bar graph, complete the following statement with one of the choices given below it. Write the letter of the correct choice in the answer space.

   In the cities in the report, air pollution inside the homes was generally ___c___

   a. greater than outside the homes.
   b. the same as outside the homes.
   c. less than outside the homes.

GOING FURTHER

6. Ventilation is a process in which fresh air from the outside is able to circulate throughout a house. Why have so many homeowners in recent years been sealing cracks around windows and openings under doors to cut down on the amount of ventilation from outside?

   If the air outside is very poluted than people won't want it inside their house.

7. Why may sealing up a house, according to the data graphed in this report, be dangerous?

   If the air inside the house is polluted it will be harder to get clean air inside.

8. More and more local and state governments are passing laws to restrict indoor smoking in public places. Why may such laws be important to all people who live or work in such areas?

   If a building is sealed + people are smoking it would be hard to circulate the air would be dangerous to coworkers

C H A P T E R
34

Interpreting a Circle Graph

✓ WORD CHECK

circle graph	a *graph*, also called a pie graph, that shows the way in which a whole quantity is divided into parts called sectors; the parts are compared in two ways: a. by comparing the fractions, percents, or other numerical data, or b. by comparing the sizes of the parts (sectors)
accidental deaths	unexpected deaths that occur without intent, by chance, or through carelessness
homicide	a killing of one human being by another
natural cause	an unexpected happening in nature causing loss or death; the death is not due to any fault or misconduct on the part of a person

PROBLEM: How do you use a circle graph to obtain information?

In a television series, a song was sung at the start of each show describing a family that, after many years of hard work, has finally gotten "a piece of the pie." Dividing something into smaller parts like the pieces of a pie is another way of expressing data in a graph—a **circle graph**.

In today's work, you will be using data contained in circle, or "pie," graphs to make comparisons and draw conclusions based on the information pictured in the graphs.

PROCEDURES

Over a 40-year span, about 50,000 people in the U.S.A. died in different types of accidents that produced five or more deaths at a time. The "pie" graph on page 141 shows the major causes of such **accidental deaths**. Use this information to answer the questions.

1. During these years, what was the most important cause of accidental deaths?

   _____fires_____

2. Which method of traveling resulted in the greatest percentage of accidental deaths?

   ___motor vehicals_____

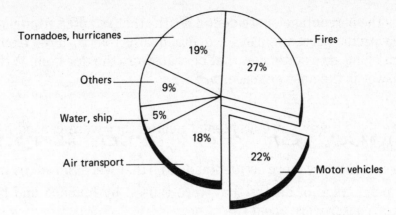

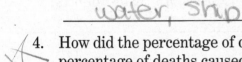

3. What method of traveling was *least* likely to result in accidental deaths?

water, ship

4. How did the percentage of deaths due to airplane crashes compare with the percentage of deaths caused by tornadoes and hurricanes? Choose a statement that best answers the question. Write the letter of the answer in the answer space.

C

   a. There were many more deaths due to airplane crashes.

   b. There were many more deaths due to tornadoes and hurricanes.

   c. The number of deaths from each of these groups was about the same.

5. The group of deaths described as "others" includes accidental deaths that took place in coal mines and on railroads. How did the percentage of deaths due to fires compare with deaths due to these "other" accidental causes? Choose a statement that best answers the question. Write the letter of the answer in the answer space.

C

   a. The deaths from both causes were about the same.

   b. The deaths caused by fires were twice as many as the deaths caused by "other."

   c. There were three times as many deaths from fires.

   d. There were four times as many deaths from fires.

6. How did deaths caused by airplane accidents compare with deaths caused by "other" types of accidents?

   #1 airplane accidents: 18% > #2 ≈ 2x #2's %

   #2 other accidentss: 9%

7. Some of the accidental deaths described resulted from the actions of humans, and others resulted from **natural causes**. Which types of accidental deaths pictured on the circle graph resulted from natural causes?

tornadoes, hurricanes, fire

8. How does the percentage of accidental deaths that resulted from natural causes compare with the percentage of accidental deaths that resulted from the actions of humans? Choose a statement that best answers the question. Write the letter of the answer in the answer space.

_____B_____

   a. There were many more accidental deaths that were caused by the actions of humans.
   b. There were many more accidental deaths that were caused by natural causes.
   c. The percentage of deaths that were caused by humans and those caused by natural causes were about the same.

9. If you were placed in charge of a program to reduce accidental deaths in the United States each year, where would you have to concentrate your greatest efforts?

   I would try to increase the safety
   of transportation
   _____

TEST YOUR UNDERSTANDING

The following circle graphs compare the causes of death in a certain year between males and females of one large U.S. city. Use the data contained in these graphs to answer the questions.

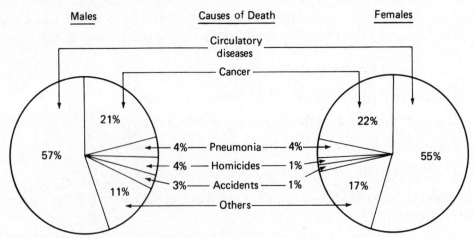

For questions 1 to 5, use one of the following phrases (a, b, or c) to complete each sentence. Write the letter of the answer in the answer space.

   a. 2% or more greater for males.
   b. 2% or more greater for females.
   c. about the same for both males and females.

1. Cancer, as a cause of death in this city in one year, was   _____C_____

2. Circulatory diseases, as a cause of death, were   _____A_____

3. The number of deaths due to pneumonia was _____ C

4. The number of deaths due to accidents was _____ A

5. **Homicides** (or murders), as a cause of death, were _____ A

GOING FURTHER

A science class took a test, and most of the students received marks between 75% and 90%. Only a few students received marks above 90%, and about an equally small percentage of the class failed. The rest of the students had marks between 65% and 74%. use this information as shown in the circle graph to answer the following questions in the Matching exercise.

Matching

Match the lettered parts of the "pie" with the statements. Write the letter of the answer in the answer space.

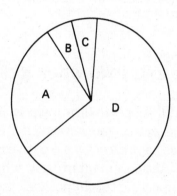

6. Which section of the circle graph represents the students who received marks between 75% and 90%? _____ D

7. Which *two* of the lettered "slices" of the circle graph illustrate the students who failed and the students who received marks better than 90%? _____ B,C

8. Which part of the circle graph describes the students with marks between 65% and 74%? _____ A

C H A P T E R
35

Testing a Hypothesis

☑ WORD CHECK

dissolves	when a solid, such as sugar, breaks up and moves throughout the water particles
precise	closely agreeing in measurements or results

PROBLEM: What happens to a solid when it dissolves in water?

When sugar is added to water, it seems to disappear as it **dissolves**. But even though you can no longer see the sugar in the water, you can check if the sugar is still there by testing the water with a chemical known as an *indicator*. A positive result from this method of testing leads to the hypothesis that substances remain present when they dissolve.

However, the trouble with supporting this hypothesis is that not all substances dissolved in water can be tested with indicators. In your work today, you will use skills you developed in previous worksheets to obtain additional evidence for the hypothesis that a substance remains present when it dissolves.

EQUIPMENT

Check off the items to make sure you have the following equipment and materials. If any item is missing, obtain it from your teacher.

___ triple-beam balance   ___ beaker   ___ stirring rod

___ 10-mL graduated   ___ sugar   ___ dropper bottle of water
      cylinder

PROCEDURES

1. Place a beaker in the center of the pan of a triple-beam balance and determine the beaker's mass. Write the result in the table in question 5.

2. Reset the balance's riders to a reading that is 1.0 g more than the mass of the beaker. (For example, if the mass of the beaker is 59.2 g, reset the riders to 60.2 g.)

3. Slowly and carefully add sugar to the beaker until the mass of the beaker and sugar reaches the new reading set on the balance. You now have exactly 1.0 g of sugar in the beaker without having had to weigh the sugar separately.

4. Remove the beaker from the balance. Use a graduated cylinder and a dropper to measure out exactly 1.0 mL of water. Add the 1 mL of water to the beaker. Stir the mixture for 30 seconds. Repeat this step (adding 1.0 mL of water to the beaker and stirring) until the sugar has been completely dissolved.

5. What would you predict should be the total mass of the beaker and its contents of 1.0 g of sugar and the added water? (Recall that the mass of 1.0 mL of water is 1.0 g.)

	Mass
Beaker	_____._____ g
Sugar	_____._____ g
Water	_____._____ g
Predicted total	_____._____ g

6. Place the beaker with the dissolved sugar and water in it back on the pan of the balance. What is the actual total mass?   _____._____ g.

7. What is the difference between the actual mass and the predicted mass?

    _____._____ g.

8. Even though the two figures for the actual and predicted mass are close, they most likely are not the same. The difference between the two figures is known as an error. Such an error or difference results when techniques and measuring equipment are not always **precise**. An error of up to 0.5 g is permitted in this experiment. What should you do if the error is greater than 0.5 g?

    _____

    _____

9. Even though the sugar is dissolved and no longer visible, why can you assume (without testing the water) that the sugar is still present?

    _____

    _____

    _____

    _____

TEST YOUR UNDERSTANDING

1. A student adds 2.0 g of substance X to a beaker that has a mass of 61.4 g. Then the student gradually adds water to the beaker and stirs the mixture until all of substance X has been dissolved. The volume of the water needed to dissolve substance X is 6.0 mL. What do you predict will be the total mass of the beaker, the dissolved substance X, and the water?

    _____._____ g.

2. After making all the measurements, the student in question 1 found that the actual total mass is 60 g. Should the student redo the measurements? Explain.

    _____

    _____

    _____

    _____

3. When iodine is added to starch, the starch changes to a blue-black color. Why is iodine solution described as being an "indicator"?

    _____

    _____

    _____

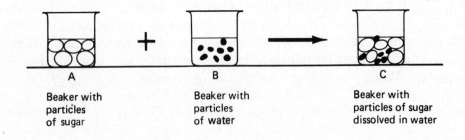

A — Beaker with particles of sugar

B — Beaker with particles of water

C — Beaker with particles of sugar dissolved in water

GOING FURTHER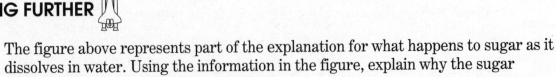

4. The figure above represents part of the explanation for what happens to sugar as it dissolves in water. Using the information in the figure, explain why the sugar particles seem to disappear in the water?

    _____

_____

_____

5. When sugar is continuously added to water, eventually a point is reached at which no more sugar will dissolve. Why is there a limit to the amount of sugar that can be dissolved in a volume of water?

    _____

    _____

    _____

6. Not all solids dissolve well in water. Why are the particles of such solids unable to dissolve?

    _____

    _____

    _____

    _____

    _____

UNIT 7
THE MICROSCOPE

C H A P T E R
36

The Microscope: Introduction

 WORD CHECK

specimen	the object being viewed through the microscope
magnify	the ability to cause specimens to appear larger than they actually are
image	a likeness of a specimen
lens	a glass or other transparent object whose oppositely facing, curved surfaces bend light rays to form an image

PROBLEM: How do you use a microscope?

While working with this book, you have learned to use a variety of different tools and measuring instruments. These are the tools and instruments used by scientists to gain information about the world around us. Now, you are about to master another basic skill, using a microscope.

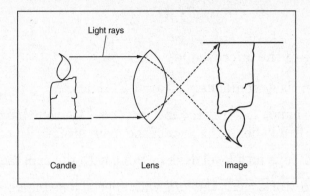

First, let us look briefly into the history of microscopes. More than 300 years ago in the Netherlands, Anton van Leeuwenhoek's hobby was grinding pieces of glass to form magnifying lenses. The figure above shows the path of light through a magnifying lens. With these lenses, he built simple microscopes, which had only one lens. His microscopes allowed him to explore a world of living and nonliving things that had never been seen before. The microscopes he constructed had the same four basic parts that are found in the more modern microscopes you will be using. See the sketch of Leeuwenhoek at work and a close-up view of one of his microscopes.

Leeuwenhoek

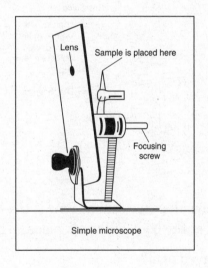

Simple microscope

The basic parts of a microscope:

1. One or more lenses that form a magnified image.

2. A place to position the specimen.

3. A source of light.

4. A mechanism to change the distance between the lens and the specimen to make the image clear. This process is called *focusing*.

EQUIPMENT

Check off the items to make sure you have the following equipment. If any item is missing, obtain it from your teacher.

_____ microscope   _____ lens paper   _____ prepared slide of the letter "e"

PROCEDURES

A. Care and handling of the microscope

1. Keep classroom aisles completely clear at all times.

2. Always use two hands to carry the microscope. Place one hand under the base while the other firmly holds the arm of the microscope.

3. Place the microscope on a level desk or table with the arm facing you.

4. Before you start work, clean the lens with special lens paper.

B. The parts of the microscope and their functions

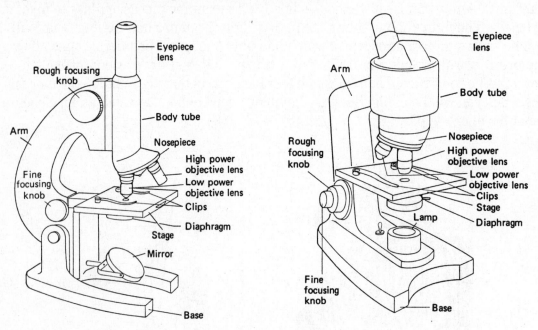

1. The **eyepiece** (or ocular) **lens** is the one nearest your eyes.
    Look at the body of the eyepiece. What number followed by an "×" do you see

    engraved on it?   _____

    This number indicates how many times this lens, by itself, would magnify the image of the specimen.

2. At the bottom of the body tube are two, or more, lenses. Since these lenses will be closest to the objects being viewed, they are called the **objective lenses**.

    How do these lenses compare with each other in length?

    _____

    What magnification does each of these lenses provide? What is the magnification of these lenses?

                                                    _____ × and _____ ×

    The shorter of the two objectives lenses is called the **low-power objective lens**. The longer lens is the **high-power objective lens**. The total magnifying power of your microscope is determined by multiplying the power of the eyepiece lens by the power of whichever objective lens is in place over the opening in the stage. To determine the total magnifying power of the microscope, multiply the power of the eyepiece lens by the power of the objective lens that is over the opening in the stage.

By how much is the image magnified, when using a 10× eyepiece lens and a

low-power 10× objective lens?   _____×

What is the magnification of the image when the 40× high-power objective lens is

used with the 10× eyepiece to look at a specimen?   _____×

3. The objective lenses are set into the **revolving nosepiece**. Leeuwenhoek could only change magnification by building another microscope. You can do it by simply positioning a different objective lens in place above the opening in the stage. Practice this with the low-power lens. Without looking through the eyepiece, how can you tell that the objective lens is now locked in place?

_____

4. A two-sided **mirror** is located between the legs of the microscope's base. One side is curved inward and the other is flat. Turn and position the mirror so that the curved surface is on top and the light coming through the classroom windows is angled up through the opening in the stage.

   CAUTION: Do not use direct sunlight.

   (Note: The procedure is just about the same if you are using an artificial light source.)

5. The **stage**, on which the specimen is placed, is located beneath the objective lenses.

   Why must the stage have a hole in it?

_____

The specimen to be observed is placed on a thin glass **slide**. The slide is held in place over the opening in the stage by the two **clips**.

6. Under the stage is the **diaphragm**, which controls the amount of light passing through the specimen. On some microscopes, the diaphragm consists of a disk with different size openings. On others, the diaphragm is a lever that can be moved from side to side.

7. There are two sets of wheels, one set larger than the other, on the sides of the arm of the microscope. Turn the larger wheels one-half turn in both directions. What happens?

_____

These are the **coarse-adjustment wheels**. They are used to bring a specimen into rough focus. Now repeat the process (no more than half a turn) with the smaller, **fine-adjustment wheels**. What seems to be happening?

_____

It is difficult to see any movement of the objective lens when using the fine-adjustment wheels because the lens moves only a microscopic distance. This movement provides a more precise focus.

C. Using the microscope

1. Position the microscope correctly on your desk or table.

2. Use lens paper to clean all the lenses.

3. Rotate the low-power objective lens until it "clicks" into place above the hole in the stage.

4. Open the diaphragm.

5. Adjust the mirror, curved surface up, or the lamp so that light shines through the opening in the stage.

6. Place the prepared slide with the letter "e" under the clips so the letter "e" is centered above the opening in the stage. The letter "e" should be right side up, as it would be if you were reading.

7. Looking from the side, gently turn the coarse-adjustment wheel so that the nosepiece moves down until you reach the "safety stop" point above the slide. This "safety stop" keeps the objective lens from being damaged or damaging the slide.

8. Put your eye to the eyepiece. (If there is only one eyepiece, try to keep both eyes open) Slowly focus upward with the coarse-adjustment wheel until the letter "e" comes into view.

9. Turn the fine-adjustment wheel for a sharper focus.

10. Draw what you see.

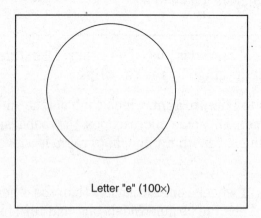

Letter "e" (100×)

11. With the specimen in focus under the low-power objective, carefully rotate the high-power objective into place above the opening in the stage.

CAUTION: Use only the fine-adjustment wheel when the high-power objective is in place. There is no "safety stop" when focusing with the high-power lens as there is with the low-power lens. Using only the fine-adjustment wheel prevents damage to the longer, high-power lens, which is closer to the stage.

TEST YOUR UNDERSTANDING

1. What is the magnifying power of a microscope that has a 15× eyepiece when a

    10× objective lens is in place?   _____×

2. Why must the specimen you are observing be very thin?

    _____

3. What should you do when viewing a specimen under high power if you completely
    lose the focus?

    _____

4. In addition to appearing much larger, how did the image of the letter "e" differ
    when viewed under the microscope?

    _____

    (The term we use to describe the changes in appearance of the specimen is to say that
    the image is **inverted**.)

5. While being observed under a microscope, some tiny living things will attempt to
    move away from the bright light. How could you adjust your microscope to reduce
    the chance of this happening?

    _____

GOING FURTHER

6. Draw the letter "P" [(placed upright on the slide)] as you would expect to see its
    image through a microscope using a 10× eyepiece and 10× objective.

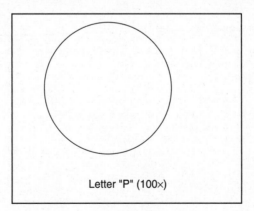

Letter "P" (100×)

7. Using a microscope with a 10× eyepiece, a student discovered that the image of
    the specimen being viewed was 200 times larger than the specimen itself. What
    must have been the magnifying power of the objective lens?

    _____ (Write the letter of the correct answer.)

         a. 10×   b. 20×   c. 30×   d. 200×

8. Arrange the microscope parts below the in correct order of the path that light travels as it goes from the mirror through the microscope.

body tube, stage, objective, diaphragm, eyepiece

_____

9. On the first page of this exercise, there is a drawing of a candle and its image. In addition to being larger than the candle itself, how does the image differ from the candle?

_____

C H A P T E R

37

The Microscope: Viewing Specimens

☑ WORD CHECK

organism	a living individual
cell	the basic unit of all living things
compound microscope	a microscope uses two magnifying lenses at the same time: an eyepiece lens and an objective lens
wet mount	the technique of preparing a specimen for observation so that it does not dry out
staining	adding a dye to a specimen to make it easier to see

PROBLEM: How do you prepare cell specimens for viewing with a microscope?

In the 1670s, Robert Hooke improved on Leeuwenhoek's simple microscope with the development of the **compound microscope**. Using this tool, he observed that a thinly sliced piece of cork from the bark of a tree seemed to be made of "many spaces resulting in the appearance of little rooms or *cellulae*." Thus we have the first use of the term *cell* to describe the tiny units that make up all living things.

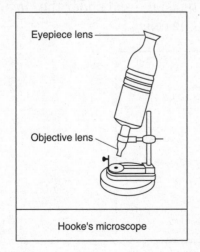

Eyepiece lens

Objective lens

Hooke's microscope

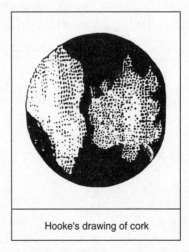

Hooke's drawing of cork

In your work today, you will add a dye to cells so that you can see them more easily. This technique is called staining. In addition, you will keep a specimen of cells from drying out by adding a fluid to the slide. This is called preparing a **wet mount**.

EQUIPMENT

Check off the items to make sure you have the following equipment and materials. If any item is missing, obtain it from your teacher.

_____ microscope	_____ coverslip
_____ lens paper	_____ cheek cells (prepared slide)
_____ forceps	_____ dropper bottle (water)
_____ slide	_____ dropper bottle (iodine)
_____ onion chunk	

PROCEDURES

A. Review the steps you should follow when preparing for microscope work (Check off each step as it is completed.)

_____ 1. Carry your microscope to your table by the arm and base.

_____ 2. Clean the lenses with special lens paper.

_____ 3. Position the arm so that it faces you.

_____ 4. Open the diaphragm.

_____ 5. Rotate the low-power objective lens into place above the opening in the stage.

_____ 6. Adjust the curved surface of the mirror or the lamp.

_____ 7. Use the coarse-adjustment wheel to lower the 10 ×, low-power, objective lens down to the safety stop.

B. Observe stained cells

1. Center the slide with the cheek cells on the stage and under the clips. This slide was prepared with a blue stain added to the cells.

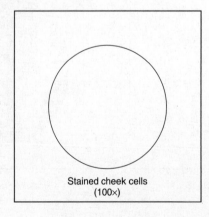

Stained cheek cells
(100×)

2. Observe the slide under low power as you slowly focus upward using the coarse adjustment wheel. Use the fine adjustment to sharpen the focus.

3. Draw three to five of the cells. Note the darker circle towards the center of each cell. This is the nucleus, which controls cell activities and cell reproduction.

C. Prepare a wet-mount slide of onion cells

1. Peel a very thin layer of cells from the inner curve of a chunk of onion. Use forceps or your fingernail to do this.

2. Prepare a wet-mount slide by placing a drop of water at the center of the slide. Place the onion cells in the drop of water. Place a coverslip over the cells to flatten them. If the sheet of cells curls up, you can unroll it with the help of pencil points.

3. Observe under low power.

    Note: It may be difficult to observe these cells because they are transparent. Closing the diaphragm almost all the way is one way to improve viewing.

4. Remove the slide from the stage of the microscope. Add one or two drops of the orange-brown iodine stain at one edge of the coverslip. Use forceps to hold a small piece of paper towel at the other edge of the coverslip to help draw the stain across the slide. As the onion cells absorb the stain they will become easier to see.

5. Draw five or six of these cells. Show their shape, arrangement with one another, and the nucleus.

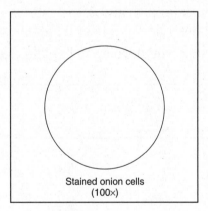

Stained onion cells
(100×)

TEST YOUR UNDERSTANDING

1. What two steps did you take to make the onion cells easier to view?

    a. _____

    b. _____

2. Why is the arrangement of the onion cells described as being bricklike?

    _____

3. Why did you peel off a single layer of onion cells rather than placing the entire chunk on the slide?

_____

4. How could you tell that what you were observing was an onion cell and not an air bubble?

_____

5. Most of the stain in the cheek cells and in the onion cells was concentrated in the nucleus of these cells. How does the nucleus differ in density from the other parts of these cells?

_____

GOING FURTHER

6. If you were trying to observe cells as they carried on their life functions, why would you not use iodine to stain them?

_____

7. Why would medical researchers be interested in developing cell stains that would be absorbed by cancer cells and not by normal cells?

_____

8. A student was observing cells taken from five different organisms and mounted on five different slides. On slides 2, 4, and 5, the cells were arranged in a bricklike pattern. The cells on slides 1 and 3 did not have this arrangement.

   a. What hypothesis could you develop to explain this difference?

_____

   b. How could you check your hypothesis?

_____

C H A P T E R

38

The Microscope: Making Measurements

✔ WORD CHECK

field	the circular area of light you see when you look through the eyepiece lens of a microscope
diameter	a straight line across the center of a circle that divides the circle in half
estimate	to make a rough approximation
millimeter (mm)	a metric unit of measurement; 1/1000th of a meter
micrometer or micron (μm)	a microscopic, metric unit of measurement; 1/1000th of a millimeter

PROBLEM: How can you estimate the size of a specimen under a microscope?

Being able to estimate the size of a microscopic specimen makes it possible to identify different types of similar organisms and draw comparisons among them. To do this, you will first have to learn how to estimate the diameter of the microscopic field and then use this information to approximate the size of the specimen.

EQUIPMENT

Check off the items to make sure you have the following equipment and materials. If any item is missing, obtain it from your teacher.

_____ microscope

_____ lens paper

_____ forceps

_____ onion chunk

_____ coverslip

_____ clear metric ruler

_____ dropper bottle (water)

_____ dropper bottle (iodine)

_____ slide

PROCEDURES

A. Estimating the diameter of the field

1. Prepare the microscope as you have done before with the 10×, low-power, objective lens in place.

2. Place a clear metric ruler on the stage. Position the ruler so that one of the millimeter lines is at the very edge of the field as shown in the diagram.

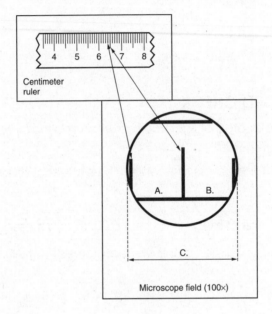

Microscope field (100×)

    a.   How wide is space A? _____ millimeter(s)

    b.   How wide is space A? _____ micron(s)

    c.   How wide is space B? _____ millimeter(s)

    d.   How wide is space B? _____ micron(s)

    e.   How wide is field C? _____ millimeter(s)

    f.   How wide is field C? _____ micron(s)

B. Estimating the size of a cell

1. Using your estimate of the diameter of the field (C), you can approximate the size of a cell.

    a.   How many cells (assume they are all about the same size) fit across the width, or diameter, of the field as shown in the diagram at the top of page 161 ? _____

    b.   Now do the arithmetic as follows

$$\frac{\text{diameter of field}}{\text{number of cells}} \rightarrow \text{number of cells} \overline{)\text{diameter of field}} \rightarrow 3\overline{)\,2000\ \mu m}^{\;?\ \mu m}$$

    c.   What is your rough estimate of the approximate size of each cell? _____ μm

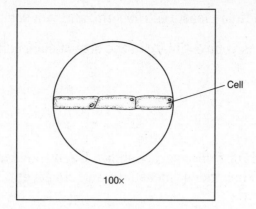

100×

C. Estimating the size of an onion cell

1. Prepare the microscope again with the 10×, low-power, objective lens in position.

2. Recheck the metric ruler and estimate the diameter of the field.   _____ μm

3. Make another wet-mount onion cell slide. But this time, place the thin sheet of cells directly into a single drop of iodine on your slide, rather than into a drop of water. Place a coverslip over the onion cells.

4. Move your slide under the clips until you have located one row of cells that forms a straight line across the diameter of the field.

5. Draw the row of cells that goes across the diameter from one side of the field to the other.

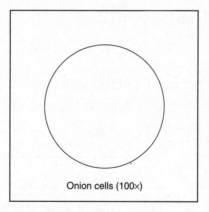

Onion cells (100×)

6. How many cells fit across the diameter of the field?   _____

7. What do you estimate the size of each onion cell to be?   _____ μm

TEST YOUR UNDERSTANDING

1. Why is it an advantage to add onion cells to iodine instead of to water?

_____

2. If the diameter of a field measures 1500 μm, and you see three cells in a line across the diameter, what is your estimate of the size of each cell? _____ μm

GOING FURTHER

A student observed cells on two separate slides. Each slide had been placed on the stage of a different microscope. The student measured and observed the following:

Slide A Diameter of the field 3000 μm	Slide B Diameter of the field 2000 μm
Number of cells reaching across the diameter ____6____	Number of cells reaching across the diameter ____4____

3. How does the estimated size of the cells on slide A compare with the estimated size of the cells on slide B?

    _____

4. If you were the student doing this work, what might you conclude about the most likely source of the cells on the two slides? _____ (Write the letter of the correct choice.)

    a.   The two sets of cells came from the same onion.
    b.   The two sets of cells came from two different onions.
    Explain your answer.

    _____

    _____

5. Why is it an advantage to be able to make measurements of cells being viewed with a microscope?

    _____

6. a.   Which one of these field diagrams on page 163 would most effectively enable you to estimate the size of the cells shown?

    _____

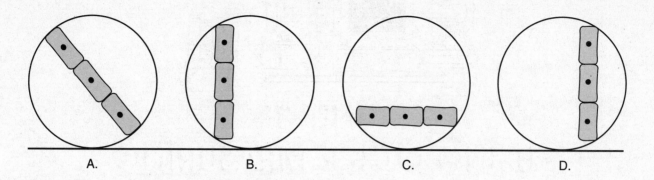

A.   B.   C.   D.

b.   Why did you chose this answer?

_____

UNIT 8
USING OTHER MEASURING
INSTRUMENTS AND TECHNIQUES

C H A P T E R
39

Using a Stopwatch

WORD CHECK

stopwatch	a hand-held instrument used to measure time
artery	a blood vessel that carries blood away from the heart
pulse	the expansion and contraction of an artery as the heart beats

PROBLEM: How can you use a stopwatch to check on a person's health?

A young child asks her mother, "Why do I have two hearts?" Puzzled, the mother replies, "What makes you think that?" "Well," the child answers, "I can feel my heart beating in my chest, but I can also feel it beating in my wrist."

To respond to her child's curiosity, the mother would have to know about the type of blood vessel called an artery. Arteries are blood vessels that carry blood away from the heart. The thick, elastic walls of the arteries expand each time the heart beats and then contract as the heart rests between beats. This alternate stretching and contracting (throbbing) of the arteries can be felt where an artery, just below the skin, crosses a harder surface, such as a bone in the wrist. This throbbing is known as the pulse. It indicates the number of times the heart beats each minute. A pulse rate that is too fast or too slow or irregular may be the result of a health problem.

EQUIPMENT

_____ stopwatch

PROCEDURES

Practice taking your pulse. Bend your wrist back, then slide the pointer finger and middle finger of your other hand down from the base of your thumb along the inner side of the wrist until you feel a throbbing under the skin. (Do not try to find your pulse with your thumb because the thumb also has a pulse.)

1. Wrist pulse rate at rest

    Sit down, relax, and rest. Your pulse rate while you are at rest is called your resting pulse rate. You will now determine your resting pulse rate. While your partner measures 30 seconds with the stopwatch, count the number of times your pulse beats during those 30 seconds. Double this number to determine your pulse rate for one minute. Then reverse the procedure: you use the stopwatch while your partner counts his/her pulse beats for 30 seconds.

    (**Note:** You can also take your pulse by placing your fingers on the side of your neck, under your jaw.)

Your Pulse Rate	Your Partner's Pulse Rate	Class Average
_____	_____	_____

2. Wrist pulse rate after exercise

    (Read step 3 before you begin step 2.)

    While you are seated, pump both your arms up and down vigorously for 30 seconds, then quickly record your pulse rate.

    CAUTIONS:

    a. Make sure your pumping arms do NOT interfere with anybody else.

    b. Do NOT participate in this part of the laboratory exercise if you have any medical condition which restricts your physical activity.

Your Pulse Rate	Your Partner's Pulse Rate	Class Average
_____	_____	_____

3. Wrist pulse rate one minute after exercise

Your Pulse Rate	Your Partner's Pulse Rate	Class Average
_____	_____	_____

TEST YOUR UNDERSTANDING

1. a. What was the highest, individual, resting pulse rate in your class? _____

   b. What was the lowest? _____

   c. What was the class average? _____

   d. How did your pulse rate compare with the class average? (circle one)

      about the same;   much higher;   much lower

   e. When trying to estimate the resting pulse rate in your class, why is it important to use the rates determined by the entire class rather than just the rate of your partner?

      _____

2. Certain factors may produce differences in pulse rates among individuals. Examples include body weight, emotional state, and age. How would you expect your pulse rate to change as you enter your classroom to take and exam? Why?

   _____

   _____

3. In general, how did the pulse rates in your class change after exercise? (circle one)

   remained about the same;   faster;   slower

4. When the body is faced with an emergency or during strenuous exercise, the bloodstream must rapidly deliver energy-producing chemicals to the muscles. What observation indicates that this is exactly what your body did?

   _____

5. One minute after the arm pumping exercise had been completed, how did your pulse rate compare with your original resting rate? _____ (Write the letter of the correct choice.)

   a. remained faster.

   b. became slower than the original rate.

   c. returned to about the original rate.

GOING FURTHER

6. When the pulse rate of an individual takes much longer than average to return to normal after exercise, it could be an indication that _____ (Write the letter of the correct choice.)
   a. the person's artery walls have hardened and lost some of their ability to stretch.
   b. the person's heart is not working as efficiently as it should.
   c. the person is not getting enough exercise on a regular basis.
   d. all of these answers are correct.

7. An EMT (Emergency Medical Technician) has been summoned to the scene of an accident. After determining that the unconscious person is breathing, the next step would be to _____ (Write the the letter of the correct choice.)
   a. give the patient a drink of water
   b. pinch the patient's nose shut and breathe into his/her mouth.
   c. pump on the patient's chest.
   d. take the patient's pulse.

   (**Note:** Never give a drink to an unconscious person. You could actually drown them by doing so.)

8. What do you think is meant by the following statement?

   "The pulse rate is the window to the heart."

   _____

9. In what other area(s) of life have you seen a stopwatch used?

   _____

CHAPTER

40

Using a Spring Scale

✓ WORD CHECK

pointer	an indicator that is attached to the spring hidden inside a spring scale
faceplate	a sheet of metal or plastic with markings on it to indicate mass or weight
range	the difference between the lowest number and the highest number written on its faceplate

PROBLEM: How do you use a spring scale to measure mass?

In Chapter 14, you learned how to use a triple-beam balance to measure mass. However, there are times when using a triple-beam balance is not practical, for example, when working outside the laboratory. Spring scales are easy to carry and easy to use because you suspend the object whose mass you want to measure from the hook on the scale.

To be precise, spring scales are used to determine weight. Weight is a measure of the pull of Earth's gravity on an object. A spring scale measures the pull of gravity on an object. When you are on Earth's surface, you can use a spring scale to determine mass, because mass and weight are related. The greater the mass of an object the greater its weight. In this investigation, you will learn how to use a spring scale to measure mass.

EQUIPMENT

Check to make sure you have a spring scale that measures grams. If you do not, obtain one from your teacher.

_____ spring scale that indicates grams (g) on its faceplate

_____ set of standard masses

168

PROCEDURES

1. Hold the spring scale by the ring at its top and allow the rest of the scale to hang straight down without touching anything. Which part of the scale's pointer is at the zero mark: the top, the bottom, or the middle of the pointer?   _____

2. What is the smallest number of grams that your spring scale can measure?

    _____ g

3. What is the largest number of grams that your spring scale can measure?

    _____ g

4. What is the spring scale's range?   _____ g

5. You will now check the accuracy of your spring scale by using it to measure the mass of some standard masses. Do NOT measure any masses beyond the range of your scale.

    5 grams   _____ g

    10 grams   _____ g

    50 grams   _____ g

    100 grams   _____ g

    200 grams   _____ g

    Did your scale's measurements agree with those found on the standard masses?

    _____

    By how many grams was it off?

    _____

    Give one reason the reading of the spring scale did not agree with the mass stamped into the standard mass.

    _____

6. Using your spring scale, measure the mass of an object that your teacher will provide.   _____ g

7. What is the range of the spring scale shown at the right?

    _____ g

8. The spring scale shown in the figure at right has some markings without numbers. For example, there are four unnumbered markings between the 0 and 50. The first marking without a number represents 10 grams and the second represents 20 grams. What number of grams does the third marking below the zero indicate?   _____ g

9. Suppose that the pointer on this same spring scale was at the second marking below the 50. Although there is no number at this position, what should be the reading?   _____ g

10. Why don't manufacturers of spring scales place numbers next to all of the markings on the faceplate?

_____

_____

TEST YOUR UNDERSTANDING

1. What is the range of the spring scale shown at right? _____ g

2. If the pointer were halfway between the marks numbered 200 and 400, what number would you record? _____ g

3. What number would you record if the pointer were on the first, very short mark below the zero (0) on the gram scale? _____ g

4. What is the scale reading shown by the pointer in the figure at the right? _____ g

5. You would find that the mass of an object would be exactly the same if you measured it on Earth or on the moon with a triple beam balance. If you used a spring scale to make the same measurements, the results would not be the same. Why not?

_____

_____

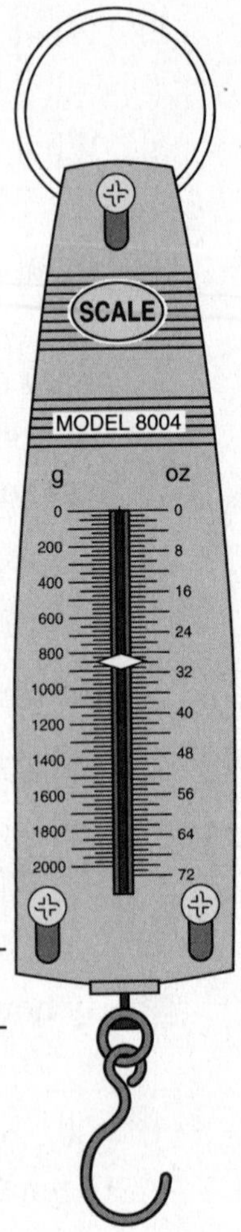

GOING FURTHER

1. Using your scale, measure the mass of a dry tea bag. Place the tea bag in a cup of water for a few minutes. Measure the mass of the wet tea bag. How do your results compare with those of other students in your class?

_____

2. Obtain different brands of tea bags. Do a science project to find out which brands soak up the greatest mass of water when tea is brewed.

41

Using a Magnetic Compass

WORD CHECK

cardinal points	the four principal directions: north, south, east, and west
compass needle	a thin magnetized piece of metal that is suspended at its center and free to rotate in a horizontal direction
true north	where the spin axis penetrates Earth's surface, at this point the latitude is exactly 90 degrees north of the equator
magnetic north	an area near Labrador, which attracts the north pole of a magnetized compass needle
face	the part below the compass needle on which the letters N, S, E, and W are printed

PROBLEM: How do you use a magnetic compass to find the cardinal points?

At one time, sailors did not travel so far out to sea that they lost sight of the shore. They used the shore and the position of the sun as references for determining their direction. At night, they could navigate by the stars. However, navigation was difficult during cloudy and stormy weather.

About the 1000s or 1100s, Chinese and Mediterranean navigators began to use magnetic compasses to determine directions. These early compasses were made with pieces of magnetic iron, which floated on straw or cork in a bowl of water. Through the years, the compass has evolved until it is the familiar pocket compass used by hikers.

This investigation should be performed in a large, open area where there are no steel beams or other large iron-containing objects nearby. You will work in groups.

EQUIPMENT

Check off the items to make sure you have the following equipment. If any item is missing, obtain it from your teacher.

_____ magnetic compass with the directions
N, S, E, W, NE, NW, SE, and SW
printed on the compass face

_____ meterstick or tape measure

PROCEDURE

1. The magnetic compass has eight symbols printed on its face. What are the full names for each of these symbols?

    N _____   S _____   E _____   W _____

    NE _____   NW _____   SE _____   SW _____

2. Stand up straight and face the direction that your teacher tells you is north. While holding the compass level in front of you, tap the compass body lightly several times. Tapping the compass reduces the effects of friction, so the needle can rotate freely. Does your compass needle point to the north when it stops moving? _____

3. Turn the case of the compass until the "N" on the compass's face is at the top and the "S" is at the bottom. Which letters are printed on the right and left sides of the face?

    Right side

    _____

    Left side

    _____

4. Observe the position of the compass needle while you are facing north. Toward what position *on Earth's surface* does the compass needle point?

    _____

5. When you take a compass reading while facing the east, toward what position *on Earth's surface* does the compass needle point? _____

6. When you take a compass reading while facing the west, toward what position *on Earth's surface* does the compass needle point? _____

7. Using the compass needle for assistance, face north. Rotate the body of the compass until the tip of the indicator needle is directly over the N on the compass face. Holding the compass tightly, make a quarter turn to your right. According to the compass, in what direction are you facing now? _____

8. Predict the compass reading if you made another quarter turn to your right.

    _____

TEST YOUR UNDERSTANDING

1. Assume that your compass's needle does *not* point to the north when the needle of the compass stops moving. Explain why the needle might point in the wrong direction.

   _____

   _____

2. A compass needle usually points to Earth's **magnetic** north pole. Under what conditions will a compass needle always point to the **true** North Pole?

   _____

   _____

3. A compass may not work properly when it is in a school building that has a great deal of iron or steel in its walls. Why?

   _____

   _____

GOING FURTHER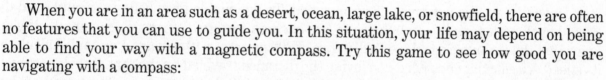

When you are in an area such as a desert, ocean, large lake, or snowfield, there are often no features that you can use to guide you. In this situation, your life may depend on being able to find your way with a magnetic compass. Try this game to see how good you are navigating with a compass:

4. While you hold the magnetic compass, have your partner place a paper bag over your head and shoulders so it also covers the compass. Stand, facing north, on a spot that you have marked on the ground. There should be enough light inside the paper bag to read the compass. If not, use a flashlight.

5. Using your compass for directions, take 10 steps to the north and stop.

6. Take 12 steps to the west and stop.

7. Take 8 steps to the south and stop.

8. Take 12 steps to the east and stop.

9. Take 2 steps to the south and stop.

10. Take off the paper bag and measure the distance between your present position and your starting point. Where are you?

    _____

11. Compare your results with those of your classmates.

    _____

    _____

42

Determining the Speed of a Moving Object

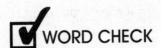 **WORD CHECK**

| speed | the distance an object covers in a given amount of time |

PROBLEM: How do you determine the speed of a moving object?

As you watch an ambulance on its way to the hospital, you know that it is moving very fast. However, you do not know its exact speed. To determine its speed, you would need to know how long it took the ambulance to travel some certain distance. For example, if you knew that the ambulance traveled a distance of 200 meters in 10 seconds, you could calculate its speed. To calculate its speed you divide the distance traveled by the time it took to travel that distance. In this case, you would divide 200 meters by 10 seconds. The speed of the ambulance would be 20 meters per second. In this investigation, you will calculate the speed of a slowly moving toy vehicle.

EQUIPMENT

Check off the items to make sure you have the following equipment and materials. If any item is missing, obtain it from your teacher.

_____ battery-operated toy vehicle

_____ 3 metersticks

_____ stop watch

_____ masking tape approximately 20 to 30 centimeters long

PROCEDURE

1. At the location assigned by your teacher, make a starting line by pasting a short length of masking tape on the floor. Also place 3 metersticks end to end on the floor.

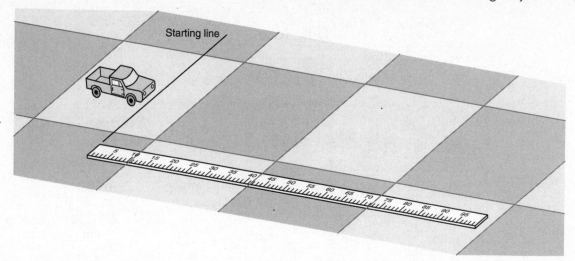

Starting line

2. To solve this problem, you will work in groups of four students. Each student will have a job to do: one will be the *engineer*, another will be the *timer*, the third will be the *data taker*, and the fourth will be the *recorder*. Switch jobs during the procedure, so that each student has a chance to perform each job.

3. Before making any measurements, try a few test runs to become familiar with the apparatus. The engineer places the toy about 10 centimeters behind the starting line. Then the engineer starts the toy and makes sure that it goes straight forward towards the starting line. When the vehicle crosses the starting line the timer starts a stopwatch. After 5 seconds, the timer calls out, "FIVE." Then without stopping the toy, after it has traveled for 10 seconds, the timer calls out "TEN" and after 15 seconds the timer calls out "FIFTEEN." Whenever the timer calls out the time, the data taker calls out the total number of centimeters that the toy has traveled from the starting line and the recorder writes the number of centimeters in the data table.

4. When the engineer, timer, and data taker are absolutely sure that they have had enough practice, they repeat the procedure for three more trials while the recorder enters the actual data in the table on page 176. Remember to switch jobs. If your data varies greatly from trial to trial, you may need to perform some additional trials.

5. Calculate the speed of the toy during the interval 0 to 5 seconds. This is done by dividing the distance from the starting line by the time interval, 5 seconds.

    _____ centimeters per second

6. Calculate the speed of the tank during the interval 0 to 10 seconds. This is done by dividing the distance from the starting line by the time interval, 10 seconds.

    _____ centimeters per second

7. Calculate the speed of the tank during the interval 0 to 15 seconds.

    _____ centimeters per second

8. On the grid provided on page 176, use your data to construct a line graph of average distance verses time. To refresh your memory on how to label and draw line graphs, you might want to refer back to Chapter 30, "Constructing a Line Graph." Draw one straight line that lies on or close to as many data points as possible.

Data Table

	Distance from starting line			
Time (t)	Trial 1	Trial 2	Trial 3	Average distance
0 seconds	0 cm	0 cm	0 cm	0 cm
5 seconds	cm	cm	cm	cm
10 seconds	cm	cm	cm	cm
15 seconds	cm	cm	cm	cm
20 seconds	cm	cm	cm	cm

TEST YOUR UNDERSTANDING

1. When should the timer start the stopwatch: the moment that the front end of the toy reaches the starting line; the middle of the toy reaches the starting line; or the rear end of the toy reaches the starting line? Explain your answer.

    _____

    _____

2. During the first 5 seconds after crossing the starting line, did your toy cover the same exact distance in each of the three trials?

    _____

    Give three reasons that could explain why the distances could be different.

    _____

    _____

    _____

3. Assume that a toy truck covered a total distance of 240 centimeters during the first 15 seconds after crossing the starting line. Predict the total distance that the truck would cover if it were allowed to travel for an additional 5 seconds.

    _____ centimeters

4. Predict the distance that this truck would go if it were allowed to travel a total of 25 seconds.

    _____ centimeters

5. Do all of the data points that you plotted on your "average distance versus time graph" lie on the same straight line?

    _____

    Explain why all of the data points should (or should not) lie on the same straight graph line.

    _____

    _____

GOING FURTHER

6. Use a string to attach a weight to the back of the toy. Predict how dragging the weight will affect the toy's speed.

    _____

    _____

7. Predict how your new average distance versus time graph will change when the toy drags the weight.

    _____

    _____

8. Repeat your original procedures and enter your new data in the table below.

Distance from starting line				
Time (t)	Trial 1	Trial 2	Trial 3	Average distance
0 seconds	0 cm	0 cm	0 cm	0 cm
5 seconds	cm	cm	cm	cm
10 seconds	cm	cm	cm	cm
15 seconds	cm	cm	cm	cm
20 seconds	cm	cm	cm	cm

9. Plot the new data for the toy on the same graph that you had drawn before. How does your new graph line differ from that of the original graph line?

_____

_____

C H A P T E R

43

Determining the Acceleration of a Moving Object

✔ WORD CHECK

| acceleration | the rate at which an object changes its speed or direction |
| incline | a surface that is tilted |

PROBLEM: How does the slope of a ramp affect the acceleration of a rolling ball?

There are many ways to change the speed of an object. If the object is stopped, you can give it a push to get it started. When an object is moving, you can put something in its path to reduce its speed or even stop it. Another way to change an object's speed it to let it roll down an incline. In this investigation, you will determine how the slope of an incline affects the speed of a rolling ball.

EQUIPMENT

Check off the items to make sure you have the following equipment and material. If any item is missing, obtain it from your teacher.

_____ golf ball

_____ 2 metersticks

_____ 2 rubber bands

_____ stopwatch

_____ 18 computer diskettes, index cards, notebooks

_____ Ping-Pong ball

PROCEDURE

1. To solve this problem, you may work in groups of two or three students. Before setting up your equipment, be sure the table is level. You can do this by slowly rolling a ball across the tabletop. If the moving ball rolls in a straight line across the tabletop, the table is level. Each student should use the equipment to record his or her own data if time permits.

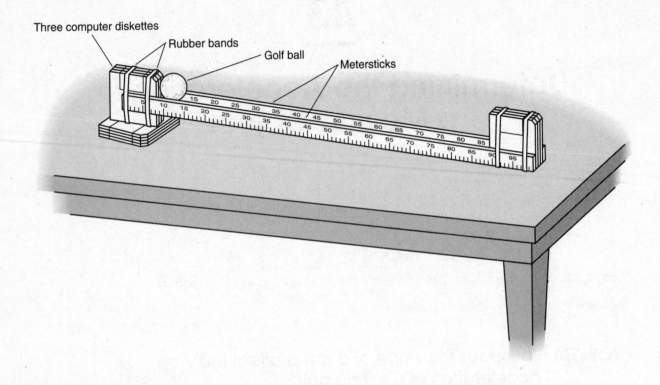

2. Bind three diskettes together with the rubber bands. Place these diskettes between the ends of the metersticks to make a ramp for the golf ball to roll down. The ball should roll along the edges of the metersticks without falling through. Check that your equipment has been assembled as shown in the diagram.

3. Record the meterstick reading under the golf ball when the ball is at the extreme left end of the ramp.

_____ cm

4. Record the golf ball's location when it is placed at the bottom of the ramp. Always measure from the same part of the golf ball.

_____ cm

5. Calculate the total distance that the ball can roll along the ramp.

_____ cm

6. Raise the left end of the ramp about one centimeter by placing a stack of three computer diskettes under it. (You may also use index cards or notebooks to raise the ramp.)

7. Release the golf ball at the top of the ramp. With the stopwatch, measure the time that it takes for the ball to reach the bottom of the ramp. In the second column, first row, of the data table on page 181, record this time to the nearest tenth of a second.

Data Table

Number of diskettes under of left end of the ramp, or height	Travel time for the golf ball	Average speed of the golf ball (total distance/ total time)
3 (1 cm)		cm/sec
6 (2 cm)		cm/sec
9 (3 cm)		cm/sec
12 (4 cm)		cm/sec

8. Perform several more trials. For each additional trial, raise the left end of the ramp with three more diskettes or by 1 centimeter. Record the travel-time data in the second column of the table.

9. For each of the trials, calculate the average speed of the ball. Enter your results in the last column of the data table.

TEST YOUR UNDERSTANDING

1. Predict the average speed of your golf ball if only one computer diskette (0.3 cm) were used to raise the left end of the ramp.

_____ cm/sec

2. Perform an experiment to see if your prediction was correct. If your prediction was not correct, write your best reason for your faulty prediction.

_____

_____

GOING FURTHER

3. If you were to use a Ping-Pong ball instead of the golf ball predict whether it would accelerate faster, slower, or at the same rate as the golf ball?

_____

4. Explain the reason for your choice in the item above.

_____

_____

5. Repeat the procedure substituting a Ping-Pong ball for the golf ball. Enter your data in the data table below.

Data Table: Using a Ping-Pong Ball

Number of diskettes under the left end of the ramp, or height	Travel time for the Ping-Pong ball (seconds)	Average speed of the Ping-Pong ball (total distance/total time)
3 (1 cm)		cm/sec
6 (2 cm)		cm/sec
9 (3 cm)		cm/sec
12 (4 cm)		cm/sec

6. Give a good reason that could explain why your prediction was, or was not, correct.

_____

_____

_____

C H A P T E R
44

Using a Voltmeter

WORD CHECK

electric potential	the energy that exists whenever there is an unequal number of positive (+) and negative (–) electric charges at a particular location
volt	the unit used to measure the difference in the electric potential between two points in an electric circuit
conductor	a metal or other material that permits the passage of an electric current
electric current	a flow of electric charges through an electric conductor
resistor	a device that opposes the passage of electric current through an electric circuit

PROBLEM: How do you use a voltmeter to measure potential differences in an electric circuit?

In 1796, Alessandro Volta wrote about his experiments to make the first electric batteries. He dipped different metals in conducting liquids to produce a small amount of electricity. In honor of his work, we now use the word "volt" as a unit of electrical measurement.

A voltmeter is used to measure the difference in the electric potential between two points in an electric circuit. You can use a voltmeter to test flashlight batteries. A fully charged battery should measure 1.2 volts.

A simple circuit is composed of a source of electric potential (the battery), wires (conductor) to carry the electric current, and something to resist the electric current (resistors). In many circuits, a switch is used to control the electric current. However, you can also control the current by connecting and disconnecting the wires in the circuit. In this investigation, you will construct a simple circuit and use a voltmeter to measure the voltage across each resistor in the circuit and the total voltage in the circuit.

EQUIPMENT

Check off the items to make sure you have the following equipment and materials. If any item is missing, obtain it from your teacher.

_____ DC voltmeter, 0-3 volt range

_____ pair of voltmeter test leads (red and black)

_____ 3 carbon resistors

_____ battery consisting of two D-size dry cells held together with a strip of duct tape

_____ bare copper wire 10 centimeters long

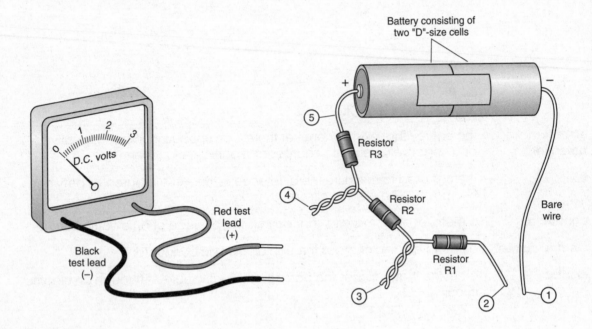

PROCEDURES

1. If it has not been done for you, use the battery, resistors, and wires to construct an electric circuit similar to the one shown in the diagram above.

   Why are the wires connected to both ends of the battery?

   _____

2. Look at the face of the voltmeter. What is the range of the voltmeter?   _____ volts

3. The wires at points 1 and 2 of the circuit act as a switch. Close the switch by touching these two wires together. While the switch remains closed, touch the black (–) voltmeter lead to point 2 and the red (+) lead to point 3. Enter your voltmeter reading (to the nearest tenth of a volt) in the second column of Data Table 1.

4. Move the two voltmeter leads to other pairs of circuit test points. Record your voltmeter readings for each pair of test points in the second column of the table.

Data Table 1

Circuit test points	Voltmeter reading (volts)
2 and 3	
3 and 4	
4 and 5	

TEST YOUR UNDERSTANDING

1. Why do the manufactures of voltmeters leave out so many numbers on the dial faces of their voltmeters ?

    _____

    _____

2. The dial on a voltmeter usually has some scale divisions that are not numbered. Assume that the needle on the voltmeter was halfway between the 0 and 1 of the dial. How many volts would that indicate? _____ volts

3. Assume that the needle on the voltmeter pointed to one small division after the zero on the dial. How many volts would that indicate? _____ volts

4. In the investigation, you were instructed to touch the black (–) voltmeter lead to point 2 and the red (+) lead to point 3. What do you think would happen if you reversed the two leads and connected the black (–) lead to point 2? Try it for a second or two and record your results below.

    _____

    _____

    _____

5. Why would it be incorrect to use a voltmeter with a range of only 2 volts for this experiment?

    _____

    _____

6. Why would it be incorrect to use a voltmeter with a range of 200 volts for this experiment?

    _____

    _____

    _____

GOING FURTHER

7. Predict what your voltmeter reading would be if you connected the leads to each of the circuit test points listed in Data Table 2. When you have finished recording your predictions, use your voltmeter and record the actual voltmeter readings in the last column of the data table.

Data Table 2

Circuit test points	Predicted voltmeter reading (volts)	Actual voltmeter reading (volts)
2 and 4		
3 and 5		
2 and 5		

8. How did your readings compare with those of other students in your class?

_____

_____

_____

45

Determining Electrical Conductivity

☑ WORD CHECK

conductivity	the ability of a material to transfer heat, light, or electrical energy
circuit	a path or a series of paths over which electricity can flow
continuity	having an uninterrupted path

PROBLEM: How can you use a simple circuit to determine electrical conductivity?

You use electrical devices every day. You listen to the radio, watch television, play CDs and DVDs, and use computers and the Internet. All these devices, whether they are plugged into outlets in the house or are battery operated, use electricity.

The device you will use today to test the electrical conductivity of different materials can also be used to test the electrical continuity of a circuit. **CAUTION:** Do Not use the tester you make to test circuits in your house. When there is a break in a circuit because a switch is open or a wire is broken, the circuit does not have electrical continuity. That is, the circuit is not complete. While it is relatively easy to find an open switch, it can be difficult to find a break in a wire, especially when it is covered by insulation. Here is where a continuity tester is useful. You can trace the path of the electricity with the tester until you find the break in the circuit.

EQUIPMENT

Check off the items to make sure you have the following equipment. If any item is missing, obtain it from your teacher.

_____ dry cell, "D" size
_____ battery holder
_____ 2 alligator clips
_____ 2 wires each about 20-cm long

_____ 1 penlight lamp miniature
     screw base, 1.2 volts
_____ miniature screw lamp base

Samples of different materials to test:

_____ aluminum _____ copper _____ cotton _____ iron _____ lead _____ plastic

_____ leather _____ paper _____ rubber _____ silk _____ wood

PROCEDURES

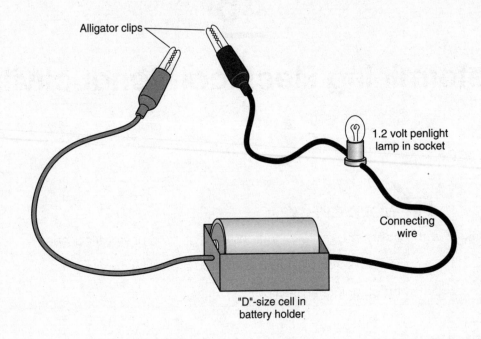

Alligator clips

1.2 volt penlight
lamp in socket

Connecting
wire

"D"-size cell in
battery holder

1. If one has not been provided, use the equipment to construct an electrical
    conductivity tester like the one shown in the diagram. Have your tester checked
    by the teacher. When your teacher has checked your tester, touch the alligator
    clips at the ends of the wires together and then separate them several times while
    observing the penlight lamp. Record your observations.

    _____

    _____

2. Connect the alligator clips to opposite ends of the piece of aluminum. Observe the
    flashlight lamp and record your observations.

    _____

    _____

3. Repeat Step 2 for each of the remaining materials. Record your observations in
    Data Table 1 on the next page.

Data Table 1

Material	Does test lamp light?
Copper	
Cotton	
Iron	
Lead	
Plastic	
Leather	
Paper	
Rubber	
Silk	
Wood	

TEST YOUR UNDERSTANDING

1. If your test lamp lights when a material is being tested, what does this indicate about the electrical conductivity of the material?

   _____

   _____

2. According to your observations, which materials make the best electrical conductors?

   _____

   _____

3. What is one important use for a material that is a good electrical conductor?

   _____

   _____

4. What is one important use for a material that is *not* a good electrical conductor?

    _____

    _____

    _____

    _____

5. Assume another group's penlight lamp does not light no matter what they try. How could you use your apparatus to help locate the source of their trouble?

    _____

    _____

GOING FURTHER

6. Some materials, such as diodes and transistors, will conduct in one direction but not in the opposite direction. Test several of these diodes for electrical conductivity and record the results of your tests below.

    _____

    _____

    _____

    _____

7. As you may already know, lead pencils used for writing contain a form of carbon, called *graphite*, rather than actual lead. Replacement "leads" for automatic pencils are available at most stationery stores. Obtain several of these replacement leads with lengths varying from 1 millimeter to 10 millimeters. Test these pieces for electrical conductivity and record your results below.

    _____

    _____

46

Measuring Angular Elevation

✓ WORD CHECK

angular elevation	the angle between a horizontal line and a sighting line to an object that is above the horizon
astrolabe	an ancient instrument that helps us measure the angular evaluation of stars and other objects visible in the sky

PROBLEM: How do you measure the angular elevation of an object with a simple astrolabe?

Before there were compasses, travelers used the North Star to guide them. Not only will the North Star tell you where north is, it can also tell you your latitude in the Northern Hemisphere. Astronomers know that Polaris, the North Star, appears to be located above the North Pole. Therefore, when an observer determines the angular elevation of Polaris, it is approximately equal to the observer's latitude. For example, a person located on the equator, which is 0 degrees latitude, would see Polaris on the horizon, or 0 degrees. A person at the North Pole would find that the angular elevation of Polaris is 90 degrees, which is the latitude of the North Pole. People living south of the equator cannot see Polaris.

In this investigation, you will construct a simple astrolabe and use it to measure the angular elevation of an object in your classroom.

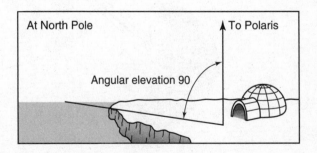

EQUIPMENT

Check off the items to make sure you have the following equipment. If any item is missing, obtain it from your teacher.

_____ protractor, large, plastic

_____ soda straw with a small hole through its side

_____ thin string, approximately 20 cm long

_____ fishing sinker (or other heavy weight)

_____ 2 small squares of duct tape (each approximately 2 cm by 2 cm)

PROCEDURES

1. If your astrolabe has not already been put together, assemble one using the equipment listed above. Thread the string attached to the weight through the hole in the straw. Tie a knot at the end of the string. Why do you need a knot at the end of the string?

_____

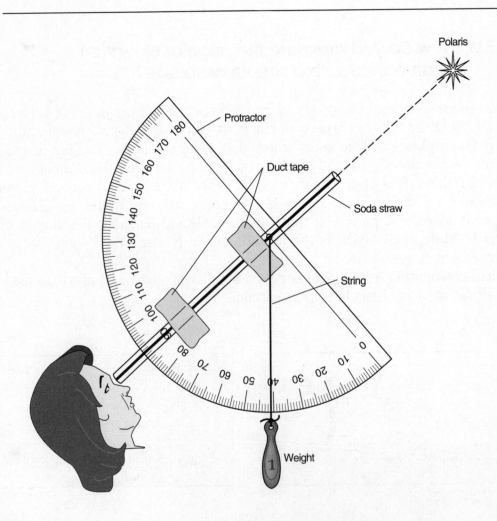

2. To assemble the astrolabe, tape a soda straw to the protractor along the 90-degree mark as shown in the diagram on page 192. Make sure the hole in the straw is on the line that connects the 0- and 180-degree marks along the flat side of the protractor. The weight should fall below the protractor's curved edge. Why must the weight fall below the edge of the protractor?

_____

3. Sit at your usual classroom seat. Sight through the straw on your astrolabe at a small star that the teacher has placed near the top of the chalkboard at the front of the room. Record the angular elevation of this star in the Data Table box that corresponds to your seat in the classroom. "Seat 1" is the closest to the chalkboard and "Seat 6" is the farthest away. When all the students in the class have made their measurement, copy their data in the appropriate boxes of the Data Table.

Data Table

Recorded Angular Elevations of the Classroom Star

	Row 1	Row 2	Row 3	Row 4	Row 5	Row 6
Seat 1						
Seat 2						
Seat 3						
Seat 4						
Seat 5						
Seat 6						

TEST YOUR UNDERSTANDING

1. Compare your data with that of the students who were sitting in different rows but at the same seat number as yours. Why might their data be different from yours?

_____

_____

2. Compare your data with that of the students who were sitting in the same row as you were. Give a reason that could explain why their data might be different from yours?

_____

_____

3. The figure on page 192 shows a student looking at the star Polaris, which has an angle of elevation of 41 degrees at the student's location. How is this 41-degree angle of elevation indicated on the astrolabe?

_____

_____

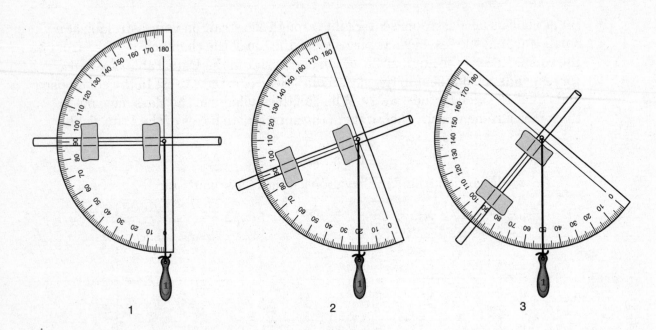

1   2   3

4. Look at the diagram that shows three astrolabes.
    a. Which one indicates a star at a 50-degree angle of elevation? _____
    b. Which one indicates a star at the horizon? _____
    c. Which one indicates a star at a 20-degree angle of elevation? _____

GOING FURTHER

5. Using a star chart, go outdoors accompanied by an adult on a clear night and observe some of the brightest stars with an astrolabe. In the data table on the next page, list the names of the stars that you saw and the angle of elevation for each one.

Name of Star	Observation Date	Observation Time	Angular Elevation
Polaris			

6. Explain why the angular elevations that you measured for a star might be different from those measured by other students in your class.

_____

_____

_____

_____

47

Separating the Parts of a Mixture

☑ WORD CHECK

mixture	a material that contains two or more substances that are not chemically combined and that are relatively easy to separate by physical means
physical properties	characteristics of a substance such as volume, density, solubility, boiling point, and magnetic susceptibility

PROBLEM: How do you separate the parts of a mixture?

In 1848, small flakes and nuggets of gold were found mixed in the sand at Sutter's Mill in California. Prospectors used a simple process called "panning" to separate the gold, which has a high density, from the sand, which has a lower density. Taking advantage of differences in density is one way to separate a mixture.

There are many other ways to separate the parts of a mixture. All methods of separating a mixture rely on differences in the physical properties of the substances that make up the mixture. For example, you can separate a mixture of zinc filings and iron filings by using the magnetic property of the iron. Iron is attracted to a magnet while zinc is not.

Crude oil is a mixture of different components such as gasoline, kerosene, and home heating oil. Oil refineries use differences in the boiling points of the components of crude oil to separate this mixture. The crude oil is heated. The substance with the lowest boiling point boils first. These vapors are captured and cooled. The mixture is now heated to a higher temperature at which the substance with a higher boiling point will vaporize.

Taking advantage of differences in solubility of the components can be used to separate other mixtures. You can separate a mixture of salt and sand by first adding water to the mixture. The salt will dissolve in the water, but the sand will not. Then, pour off the water and let the water boil away, leaving the salt behind.

In this investigation, you will learn how to pan for gold and apply other methods of separating mixtures.

EQUIPMENT

Check off the items to make sure you have the following equipment. If any item is missing, obtain it from your teacher.

_____ gold panning pan or frying pan

_____ magnifying glass

_____ package of sand (approximately 300 grams) containing a few small gold flakes or nuggets

_____ 1 liter of water in a beaker or other suitable container

_____ basin or large pail

PROCEDURES

1. Pour the contents of the sand package into the pan provided by your teacher. Shake the pan from side to side until the surface of the sand is flat. Use the magnifying glass to look for gold particles lying on the surface of the sand. Record your observations.

_____

_____

2. Add water to the pan until the water is about 3 centimeters deep. Hold the pan over a basin to catch any water that spills. Shake the pan to agitate the sand. Would any gold particles mixed with the sand rise to the top of the sand or fall to the bottom?

———————————————————————————————

———————————————————————————————

3. Tilt the pan to spill most of the water and some of the sand into your basin. Then add some more water. Holding the pan flat, swirl the pan in a half circle and suddenly stop, allowing some of the sand and water to spill into the basin. Continue this process until only a teaspoon of sand remains in the bottom of the pan. How many pieces of gold are you able to see with your magnifying glass now?

———————————————

4. If you did not find gold hidden in the sand, do not give up yet. Pour all of the discarded sand from the basin back into the pan and repeat all of the steps above. Record your results.

———————————————————————————————

———————————————————————————————

TEST YOUR UNDERSTANDING

1. Why would you expect to find particles of gold near the bottom of the sand rather than at the top surface?

———————————————————————————————

———————————————————————————————

2. When panning, why is it necessary to swirl the sand and water and then stop suddenly?

———————————————————————————————

———————————————————————————————

3. Why are gold-prospecting pans usually made of a lightweight, black plastic rather than a shiny metal?

———————————————————————————————

———————————————————————————————

4. Which is heavier, a grain of sand or a gold nugget of the same size?   ——————————

Explain your choice.

———————————————————————————————

———————————————————————————————

GOING FURTHER

5. Why is it very difficult to separate a mixture of iron filings and dark sand by panning?

    _____

    _____

6. Write a procedure that could be used to separate such a mixture.

    _____

    _____

    _____

7. How could you separate a mixture of liquids, when each of the liquids has a different boiling point?

    _____

    _____

    _____

8. Write a procedure that you could use to separate a mixture of salt and water.

    _____

    _____

    _____

Glossary

acceleration the rate at which an object changes its speed or direction.

accidental deaths unexpected deaths that occur without intent, by chance, or through carelessness

acid a substance, such as vinegar, that turns blue litmus red

agar a jellylike material that contains nutrients for growth of bacteria studied in the laboratory

AIDS (acquired immunodeficiency syndrome) the group of diseases or conditions resulting from an invasion of the body by the HIV virus

AIDS virus a virus (Acquired Immune Deficiency Syndrome) that destroys one of the body's important defenses against diseases

air pollution the contamination of the air, especially with waste materials produced by the human population

angular elevation the angle between a horizontal line and a sighting line to an object that is above the horizon

antibodies chemicals in the blood that work with T cells to destroy viruses

Archimedes a Greek mathematician, scientist, and inventor who lived during the third century B.C.

artery a blood vessel that carries blood away from the heart

astrolabe an ancient instrument that helps us measure the angular elevation of stars and other objects visible in the sky

atmosphere the whole mass of air surrounding Earth or the air of a specific area (locality)

bacteria one-celled organisms that cause disease

bar graph a graph in which numerical data are shown as bars of different lengths

base a substance, such as soap, that turns red litmus blue

boiling point the temperature at which a liquid begins to boil

Bunsen burner a heating device that mixes gas and air to provide a very hot flame for laboratory use

carbon dioxide a type of gas found in fire extinguishers

cardinal points the four principal directions: north, south, east, and west

cell the basic, life-functioning unit of all living things

centimeter one one-hundredth of a meter

chemicals substances used in experiments

circle graph a graph in which numerical facts are represented by parts (sectors) of a circle

circuit a path or a series of paths over which electricity can flow

circulatory diseases diseases that affect the system of blood, blood vessels, lymphatics, and heart of the human body

clips devices that hold the slide in place on the stage of a microscope

coarse-adjustment wheel used to bring a specimen into rough focus

compass needle a thin magnetized piece of metal that is suspended at its center and free to rotate in a horizontal direction

compound microscope uses two magnifying lenses at the same time: an eyepiece lens and an objective lens

conductivity the ability of a material to transfer heat, light, or electrical energy

conductor a metal or other material that permits the passage of an electric current

contaminated unfit for use; spoiled

contraction a decrease in the volume or original size of an object

continuity having an uninterrupted path

control experiments that are usually conducted in duplicate when testing a hypothesis and that lack a procedure, agent, or chemical under test; a standard of comparison in judging experimental outcomes

coverslip thin piece of glass placed over the specimen on a slide

crucible a porcelain container in which to heat substances that require a high degree of heat to melt

cube a regular solid with six equal square sizes

cubic centimeter the volume of a cube whose edge is one centimeter

decade a period of ten years

density mass per unit volume

diagnosis the identification of a disease based on the symptoms

diameter a line that divides a circle in half

diaphragm regulates the amount of light passing through the specimen on the stage of a microscope

dimension a measure in one direction, such as the length, height, or width of an object

dissolve to cause a solid, such as sugar, to break up and move throughout a liquid, such as water

electric current a flow of electric charges through an electric conductor

electric potential the potential energy that exists whenever there is an unequal number of positive (+) and negative (−) electric charges at a particular location

epidemic a sudden and rapid spread of a disease among many individuals at the same time

error the difference between an observed or calculated value and the true value

estimate a rough or approximate calculation of something's value

expansion an increase in the volume, or original size of an object

extinguish to cause to stop burning; to put out a fire

eyepiece lens (or ocular) the lens of a microscope nearest your eyes

face the part below the compass needle on which the letters N, S, E, and W are printed

faceplate a sheet of metal or plastic with markings on it to indicate mass or weight

fact something that is known to be true

field the circle of light seen through the microscope

fine-adjustment wheel used to bring a specimen into sharp focus

fire prevention measures that are taken to keep harmful fires from starting

flask a liquid container with a long, narrow neck

focusing changing the distance between the specimen and the objective lens

freezing point the temperature at which a liquid begins to freeze

goggles protective glasses worn when conducting experiments

graduated cylinder an instrument used to measure the volume of a liquid

gram a unit of mass in the metric system

gravity includes the force of attraction between the earth or moon and objects on their surface

helper T cells a specialized type of white blood cell that helps to destroy viruses

high-power objective lens the longer objective lens on a microscope; it provides greater magnification

HIV (human immunodeficiency virus) the virus that cripples the body's immune system by destroying helper T cells

homicide the killing or murder of one human being by another

Hooke developed the compound microscope, and was the first to use the term "cell"

host cells   cells that, when invaded by viruses, are forced to produce new viruses

hot plate   a piece of equipment used to heat substances without using an open flame

hypothesis   an educated guess that is usually based on information already collected and that explains a situation, condition, or operation in nature

identifying property   a unique or distinguishing characteristic that serves in identifying an unknown substance

image   the likeness of a specimen

incline   a surface that is tilted

indicators   chemicals which show something is present by changing their color

information   one or more facts that are known

insulated   made of a material that prevents a loss of heat to surroundings

interval   a space between two lines or points on the scale of a measuring instrument; a space of time between events

inverted image   describes the right-to-left, top-to-bottom reversal produced by light passing through a lens

irregularly shaped   having uneven surfaces rather than even sides

Leeuwenhoek   early developer of the microscope (1600s)

lens   a glass or other transparent substance that produces an image

life expectancy   the number of years an average person born in a certain year is expected to live

line graph   a graph in which numerical data are represented by points with connecting line segments

liquid   one physical form of a substance in which the particles of the substance are held closer together than they are in the form of a gas

liter   unit of volume; 1000 cm³ or 1000 mL

low-power objective lens   shorter of the two objectives lenses on a microscope, it provides lower magnification

magnetic north   an area near Labrador, which attracts the north pole of a magnetized compass needle

magnification   the number of times the specimen is enlarged

magnify   to enlarge

mass   a measure of the quantity of matter in an object or thing

matter   anything that occupies space and has mass

meniscus   the curved upper surface of a liquid column

meter   a unit of length in the metric system

micron   (μ) microscopic unit of measurement; 1/1000th of a millimeter

milliliter   (mL) one one-thousandth of a liter

millimeter   (mm) metric unit of measurement equal to 1/1000th of a meter

mixture   a material that contains two or more substances that are not chemically combined and are relatively easy to separate by physical means

mold   an organism that cannot make its own food and that lives on other living things or on materials

natural cause   an unexpected happening in nature causing loss or death

neutral   neither acid nor base

nucleus   controls cell activities and cell reproduction

objective lens   magnifying lens located closest to the objects being viewed with a microscope

organism   a living thing

particle   a very small portion or amount of a substance

petri dish   a special flat container with a cover that is used for growing bacteria in the laboratory

physical property   a characteristic of a substance such as volume, density, solubility, boiling point, and magnetic susceptibility

physical quantity   something in our universe that is measurable, such as length or time

plant cells   typically arranged in a bricklike pattern

pointer   an indicator that is attached to the spring hidden inside a spring scale

precise   closely agreeing in measurements or results

problem   a question that needs to be answered or solved

properties   identifying characteristics of a substance, such as boiling and freezing points

pulse   the expansion and contraction of an artery as the heart beats

range   the difference between the highest and lowest number of units that a device can measure

regularly shaped   an object formed or built according to some established rule, standard, or pattern

resistor   a device that opposes the passage of electric current through an electric circuit

revolving nosepiece   the objective lenses are attached here

rigid   appearing stiff or firmly inflexible

safe   free from the threat of danger, harm, or loss

scale   a series of lines or points that mark off known spaces, intervals, or distances on a measuring instrument

simple microscope   only one magnifying lens

slide   the thin piece of glass on which is placed a specimen to be viewed with a microscope

solution   the answer to a problem

solve   to find the answer to a problem

specimen   the object being viewed through the a microscope

speed   the distance an object covers in a given amount of time

spore   the reproductive cell of a mold

square   a rectangle with all four sides equal

stage   the area of a microscope on which the specimen is placed

staining   the technique of adding a dye to a specimen to make it easier to see

standard unit of measure   an established rule for the measure of a physical quantity

stopwatch   hand-held instrument used to measure time

Styrofoam   a rigid plastic material often used as an insulator

substance   material that makes up an object or thing

surface tension   the skinlike property of the water's surface

symptoms   signs of a disease

tallying   making a count of items by checking them off

temperature   how hot or cold something is as measured with a thermometer

thermometer   an instrument that has a glass bulb and a tube which contain a liquid that is sealed in and rises and falls with changes of temperature

thistle tube   a funnel with a bulging top

true north   where the spin axis penetrates Earth's surface, at this point the latitude is exactly 90 degrees north of the equator

unique   being the only one

viruses   extremely small particles that can cause diseases by entering cells, multiplying, and destroying the cells

volt   the unit used to measure the difference in the electric potential between two points in an electric circuit

volume   the amount of space or cubic units that a substance occupies

weight   the measure of the earth's gravitational attraction for an object

wet mount   placing a specimen on a slide with a fluid to keep the specimen moist